How Your Brain Works

How Your Brain Works

Neuroscience Experiments for Everyone

Greg Gage and Tim Marzullo

The MIT Press

Cambridge, Massachusetts | London, England

The MIT Press would like to thank the anonymous peer reviewers who provided comments on drafts of this book. The generous work of academic experts is essential for establishing the authority and quality of our publications. We acknowledge with gratitude the contributions of these otherwise uncredited readers.

This book was set in ITC Stone Serif Std and ITC Stone Sans Std by New Best-set Typesetters Ltd. Printed and bound in the United States of America.

Cover design by Derek George with illustrations by Matteo Farinella.

Library of Congress Cataloging-in-Publication Data

Names: Gage, Greg, author. | Marzullo, Tim, author.
Title: How your brain works : neuroscience experiments for everyone / Greg Gage and Tim Marzullo.
Description: Cambridge, Massachusetts : The MIT Press, [2022] | Includes index.
Identifiers: LCCN 2021060556 (print) | LCCN 2021060557 (ebook) | ISBN 9780262544382 (paperback) | ISBN 9780262371278 (pdf) | ISBN 9780262371285 (epub)
Subjects: LCSH: Neurosciences—Experiments. | Neurosciences—Popular works.
Classification: LCC RC337 .G34 2022 (print) | LCC RC337 (ebook) | DDC 616.8—dc23/eng/20220217
LC record available at https://lccn.loc.gov/2021060556
LC ebook record available at https://lccn.loc.gov/2021060557

10 9 8 7 6 5 4 3 2

How the Brain Works

TABLE of CONTENTS

Contents

Part III: Systems Neuroscience

INTRODUCTION

1
Introduction

Why DIY Neuroscience?

We breathe, and we see our chest rise and fall. We walk in hot weather, and we sweat. We move our arms and legs, and we see the muscles contracting below our skin. We become excited when approaching someone we have a crush on, and we feel our heart rate increase. We can remember the smell of our grandmother's basement, what our first kiss felt like, and our home address. These are all possible due to the wonderful organ inside our head called the brain. Understanding the brain remains one of the greatest scientific challenges.

How does thinking actually occur? How does your brain tell your body to move? How does your body tell your brain about its multiple senses? How do we remember? Why do we dream? How are we conscious and aware of ourselves? How do we learn? These questions have perplexed thinkers since early civilization and have evolved into the field of neuroscience that seeks answers. For the past 150 years, great progress has been made in understanding brain function. However, typically only neuroscientists have appreciated these findings. Unlike earth sciences, plant biology, physics, astronomy, and other mainstays of the education system, neuroscience has not been traditionally taught until advanced studies at a university.

Wait . . . Neuroscience Is Hard!

Neuroscience is perceived to be too complex or too expansive to learn in high schools or to take on yourself. Adages and truisms such as "it's not brain surgery," or "it's not rocket science" imply that anything having to do with brains or rockets is too cognitively difficult. The implication is that only a select few can even try to tackle these subjects at research universities. These phrases may also pertain to high-risk situations—brain surgery is dangerous and can cause damage to people, and rockets can blow up. Perhaps it is not surprising that neuroscience is typically taught at the university level, and experiments using living brains are often only conducted at well-funded research institutions.

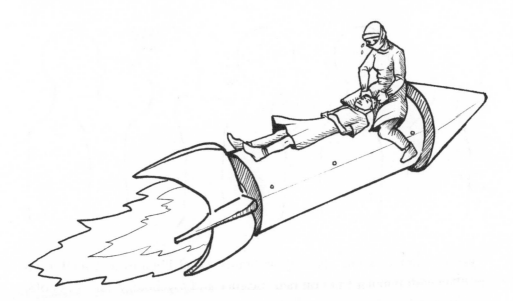

But perhaps a better reason why neuroscience doesn't find its way into more classrooms has nothing to do with our brain's complexity, but rather because the research tools are too expensive. There are a limited number of active neuroscience researchers when compared to typical consumers, so the market that has developed to supply researchers with equipment must charge a premium to stay in business. This is not a problem for well-established neuroscientists as they budget for this in their grants, but it makes neuroscience tools out of reach for most high schools and colleges.

The Growing Need for Neuroscience Education
While we have made great strides in understanding the brain, we are still in the medieval times in the broader field of neuroscience. We still do not know

exactly how memory is stored in the brain. The medical community cannot reliably diagnose Alzheimer's disease until the brain is sliced after death. What is schizophrenia, exactly? Or depression? One out of five people will be diagnosed with a brain affliction at some point in their lives, and we notoriously have no cures for neurological disorders. Basic and accessible brain research is needed to change this.

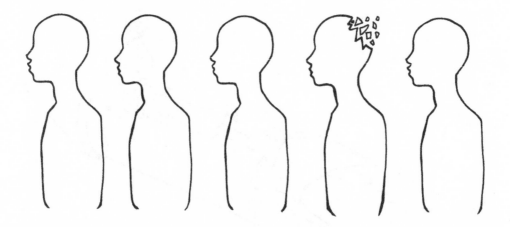

People want to know about their brains. Local libraries and bookstores are filled with popular titles on neuroscience and psychology. Piles of books, written by talented writers, philosophers, neuroscientists, psychologists, and computer engineers, attempt to explain the brain in ways that are ever more engaging. Magazine covers often trumpet stories about the "mysteries of the brain" revealed. This fascination with the brain speaks to our thirst for knowledge. The brain is personal, it is mysterious, it governs our entire life. It even seeks to understand itself.

Unfortunately, the lack of neuroscience education leaves the field wide open for unneeded exploitation. Fields like neuroeducation, neuromarketing, and neuroeconomics often benefit from a lack of public understanding of basic brain function. There is a growing market for brain health improvement products: vitamin supplements, omega oils, and apps that play classical music to your kids. There are also common brain misconceptions that permeate society. Ideas like "you only use 10% of your brain" may inspire us to work harder;

"alcohol kills brain cells" may help you drink less; or "crossword puzzles keep your brain young" may keep you engaged and occupied; but none of these claims are based on neuroscience research. There is no physical evidence for "left-brained" or "right-brained" personalities, but these myths demonstrate an underlying interest in the brain and its role in our behaviors.

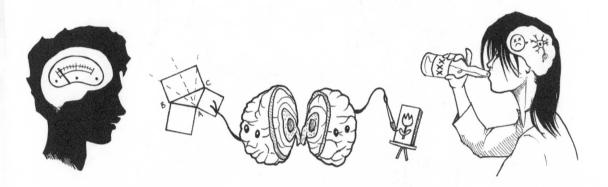

Neurorevolution Is Near

The interest in the brain paired with the lack of consumer neuroscience research gear resemble the early days of the computer revolution. Computers were available even in the 1960s, but they were large and expensive. Only banks, businesses, and large universities could afford them. However, with the invention of the Intel 8080 microprocessor, there was finally a chip affordable enough ($75–$300 in 1976) to introduce the first personal computer, the Altair 8800. This was the spark that ignited the computer revolution. Communities of hobbyists soon formed and ideas were shared. Steve Wozniak credits the Homebrew Computer Club as the inspiration for the Apple I. Microsoft's first product was a BASIC language interpreter for the Altair 8800. The acceleration of the personal-computer innovation started with the democratization of microprocessors and continues to this day.

The electronic/computer revolution has had the "spillover effect" whereby, with the invention of low-cost components and open-source tools, it is now possible to bring a wider population to the field of neuroscience by providing the electronics needed to record the electrical activity of brain cells.

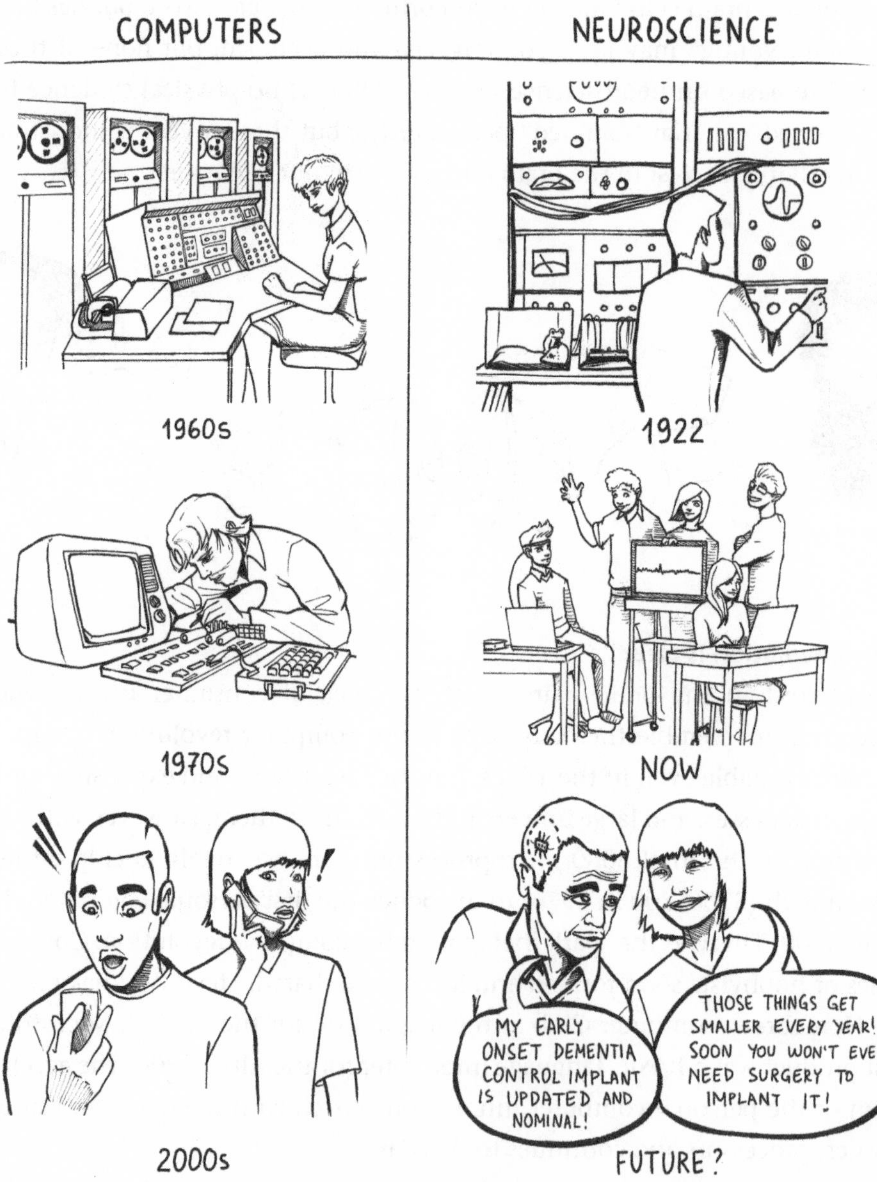
COMPUTERS
NEUROSCIENCE
1960s
1922
1970s
NOW
2000s
!
MY EARLY ONSET DEMENTIA CONTROL IMPLANT IS UPDATED AND NOMINAL!
THOSE THINGS GET SMALLER EVERY YEAR! SOON YOU WON'T EVEN NEED SURGERY TO IMPLANT IT!
FUTURE?

The goal of DIY neuroscience is to replicate the computer revolution by allowing people who previously did not have access to the tools to contribute to the field in meaningful ways.

There are many examples of how amateur scientists add to our collective understanding of nature. As soon as astronomy clubs and groups began to spring up, amateurs began discovering new nebulae and comets in the night sky. In fact, the famous Hale-Bopp comet was codiscovered by an amateur astronomer Thomas Bopp. Even the planet Uranus was discovered in 1781 by a curious music director named William Herschel. Many rare occurrences such as comets and large asteroids crashing into planets would have gone unnoticed if not for the watchful amateurs who preserved these events on video for scientists to study.

In mathematics, some of the most elegant proofs for calculating *pi*, amid many other astounding derivations, came from an amateur in India named Srinivasa Ramanujan. Pet owners have greatly contributed to animal cognition work. For example, Irena Schulz's cockatoo Snowball lead to the discovery that non-humans can develop beat induction (matching the rhythm of music), and Rico, a dog from Germany, lead to a 2004 publication in the journal *Science* that described his ability to classify up to 200 novel objects.

One thing these amateur scientists all have in common is access to inexpensive tools. Telescopes, pencils, and pets are affordable, so data and ideas can easily be generated. If neuroscience tools could be democratized in a similar way, perhaps amateurs could help us improve our understanding of the brain as well.

Fortunately, in 2008 several disruptive consumer technologies were released that changed the way that traditionally expensive tools could be developed. The Arduino microcontroller was invented by a group of artists and engineers to allow amateurs to easily control electronic and mechanical devices, leading to inventions like the MakerBot, the first commercial 3D printer for consumers. That same year, Apple released the App Store for their iPhone mobile device, turning the iPhone into an incredibly powerful internet-connected portable computer that users could program. There were now the raw materials to build parts, program electronics, and connect devices to mobile apps. These technologies led to a new type of community center called a makerspace. Desktop-manufacturing tools could be shared and allowed almost anyone to assemble a prototype of anything: a rocket, a new musical instrument, even scientific instrumentation.

Using these DIY techniques, it is now possible to build simple neuroscience tools that take the place of advanced laboratory equipment. Experiments from homemade neuroscience gear can be wide-ranging. Neuroscience is big and small at the same time. You can zoom in and eavesdrop on the electrical messages of single neurons communicating with each other in insect legs, or zoom out all the way up to the broad electrical activity of billions of neurons in your own brain through electroencephalography (EEG). The field of neuroscience is as diverse as the people who study it. The annual Society for Neuroscience meeting, the largest conference in neuroscience, has about 30,000 people in attendance each year. This sounds like a lot, but it is nothing but a small town. With the democratization of tools and skills, as well as picking the right questions to ask, it is possible for many more to contribute to the field.

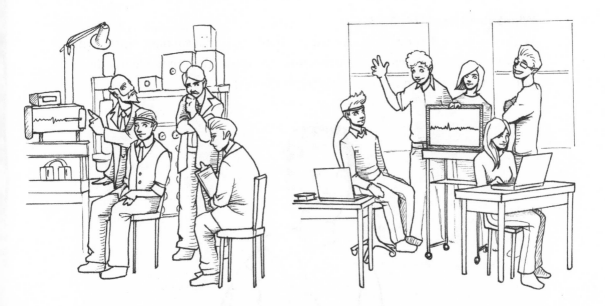

This book allows you to learn about neuroscience research and actively participate in the field. We aim to provide access to tools and techniques so you can develop a sense for what questions to ask using neuroscience experiments and what the limits are of current technology. What was previously only found in labs at advanced research universities is now available for anyone with curiosity. The walls have disappeared, the locks are gone, the bars have been lifted, and the drawbridge has lowered. Neuroscience and self-knowledge of your brain are yours for the taking.

How to Use This Book

This book is a practical guide to learning about the brain through hands-on, tactile experiences. We will introduce concepts in neuroscience, a bit of the history of where these come from, and then walk through a number of experiments that *you* can do to understand the brain a bit more. By interacting with neuroscience directly, you will understand the basic principles that will begin to demystify how your brain works.

Who Should Read This Book

While we think that everyone is curious about the brain, we have written this book with several specific audiences in mind. While they may seem a disparate lot, the interest in neuroscience binds them all.

Students If you are a student of biology, physiology, or anatomy, you are often left with little lab experience around neuroscience. This book will complement and broaden the general knowledge you've acquired. You may want to replicate the experiments or build the circuits shown in this book to get an applied knowledge of neuroscience and biomedical engineering. This will give you a "boots on the ground" view of what neuroscience is in practice and may guide you in future career choices. This book can also serve as a springboard to think about science projects in neuroscience for middle school science fairs or university senior thesis projects.

What you do with this knowledge is up to you. As an analogy, let's say you are thinking about pursuing music. One way to do this is to buy a low-cost keyboard or guitar and begin learning how to play it. Your new hobby may stay a hobby, or it may develop into a serious passion that will bump up your life to another level—like forming a band with your friends. It might even blossom into a full-blown career in music. Whatever the outcome, your life will be a little bit better because of music. Our goal is to do the same for neuroscience. You can use the book to learn enough about neuroscience to be an informed citizen or to become a serious amateur (like astronomers and chess players), or you can use this book as your inspiration to become a professional neuroscientist, biomedical engineer, or physician (or all three, depending on your ambition).

Parents   If you are a parent who wants to encourage an interest in biology in your child, this book is for you too. Your child may not even know that studying the electrical systems of living things can be satisfying and immensely rewarding, full of mystery and wonder. You can help guide and nurture the capacities of your child by working together on projects. Neuroscience is a very broad field, and there is still much left to be discovered. A young student can contribute to the field of invertebrate neurobiology, and nothing is more attractive to a college admissions board than examples of independent research and thinking, especially in high technology. Our content aligns with the Next Generation Science Standards (NGSS) and AP Biology Framework, so your child will begin to grasp advanced fundamentals through the concept of exploratory play.

Learning about the brain is also fun. Not only do children have brains, but adults do as well. So, do the experiments together and learn together! Doing projects jointly can create some of the strongest and enduring memories between parents and children (and between friends). A child may hold onto a tool they used together with a parent or grandparent until late into adult life, until it is time to pass it onto the next generation.

Curious Souls   If you are someone who likes to read popular science books written from an instructional and practical view, then this book is for you. Here you will learn about the brain in an applied way. We will not use long paragraphs filled with words that describe anatomical structures and connections that you have never heard of. Instead, we will focus on a first-principles approach to cutting-edge neuroscience. We will demonstrate how to build measuring equipment to study the brain, and how to work with these tools using illustrations and analogies along the way.

We too are inspired by the DIY resources found online or in print in magazines like *Make*. They provide detailed instructions to replicate projects developed by artists, engineers, amateurs, and scientists. We have learned a lot through replication and want to apply the same nitty-gritty, hobbyist, nuts-and-bolts detail to popular neuroscience. This has been lacking in neuroscience books to date. This book will help you become a more biologically informed citizen. We want you to cringe the next time you read a popular press article about a robot or video game being controlled by brain wave activity. You will see firsthand the limitations of various neuroscience technologies, and you will be able to collect the data yourself. The more you understand your brain and your biological signals, the less likely you are to believe outrageous claims. Armed with a little curiosity and the tools to answer your questions, you may improve the field of neuroscience along the way!

Scientists   If you are a professional scientist considering bringing electrophysiology into your research or teaching program, this book is for you. You can treat this as a practical reference book. Neuroscience has many sub-disciplines, and one often cannot have practical knowledge in many more than one or two. Our goal is to bring those outside of electrophysiology into the fold. Many scientists have a copy of Horowitz and Hill's *The Art of Electronics* or Halliday, Resnick, and Walker's *Fundamentals of Physics* on their bookshelves to help with engineering and physics questions. Neuroscientists may also have a copy of *Campbell Biology* as a reference for biological processes. Eric Kandel's *Principles of Neural Science* is an excellent reference for the entire field of neuroscience. In fact, neuroscience is a field so broad that Kandel's book weighs more than the brain itself! We aim for this book to be your on-the-shelf, go-to resource for anything electrophysiology or biosignal related.

What are the differences between neural signals and heart signals? Where should electrodes go when recording brain rhythms? What is the difference between a wave and an evoked potential? Our book aims to help you on these questions, with chapters that try to explain the basics using the most economical language we can devise. You can get up to speed in electrophysiology within a week.

Navigating the Book

How you read this book is up to you. You can read this book from cover to cover from the comfort of your kitchen table with a coffee in hand to learn more about neuroscience investigation techniques, or you can follow along and do all the experiments yourself. The sections are grouped into orders of neural complexity, from a small number of nerve cells to how groups of these cells work together in the nervous system. This does not mean that the subject gets more complex. While learning about neurons is a good starting point, you could readily begin your inquisition with human muscle physiology or electrocardiograms (EKGs). The principles you will learn from one system are often relatable to others.

Chapters begin describing some phenomenology of a particular area of neuroscience. For example, why are you only aware of background sounds like your refrigerator motor only after the power goes out? Instead of discussing the theoretical background at length, we set the stage for the main event: the experiment. Here is where you get to actually measure things. We will describe how to set up and run neuroscience experiments on your own. We will show data from our experiments and discuss the results with you. It's important to note that anytime you see a data trace, it comes from real data—not an artist's brush. Some data is corrupted by a little noise, others by EKG artifacts, but this will be similar to what you would record at home. The experiments will guide you through a series of questions, as we build up some theory and even history. We kept the titles of the experiments as they appear in the scientific literature to allow you to find and compare your results with fellow scientists in the field. But the science doesn't end there. Each chapter ends with open-ended questions for experiments that you can run on your own. (For example, do reaction times get longer with age?) We will follow this structure with every chapter.

The point of this book is to do experiments! Therefore, we designed this book so that it will stay open when placed on the table so your hands can be free to work. The DIY scientific tools mentioned in the book are commercially available from our organization, Backyard Brains, but they can also be built by you from open-source schematics and software. We have added appendices to show how you can build the bioamplifiers completely from scratch using breadboards and op-amps, and how to care for pet invertebrates. Regardless of your approach to this book, at the end you will be practiced and knowledgeable in advanced neuroscience techniques. Whether you are a motivated student, a nurturing parent, a curious citizen, or a professional scientist, your road to discovery awaits.

The Ethics of Using Animals

In this book, we will be recording neural signals not only from us humans but from invertebrate animals as well. Whenever we work with animals in science, it is important to first have an ethical discussion about the use of animals. While many believe there is no middle ground when it comes to using animals in science, scientists have developed an ethical framework that can help guide decisions about what should be permitted and not permitted when working with animals.

The relationship between animals and humans is complex and goes back at least 12,500 years when dogs became the first domesticated animals (followed by goats and sheep), and, of course, our hominid ancestors have hunted animals for food for at least 2 million years. In our modern era, there are many uses of animals in society and many associated ethical debates. Opinions can range from those who believe that animals should have the same rights and protections as human beings to those who endorse that a responsible use of animals should meet certain human needs.

We will first discuss examples of the main uses of animals in society as well as the level of controversy of each use. We can then look at how an ethical framework could be used to evaluate these scenarios, to help guide us in making decisions.

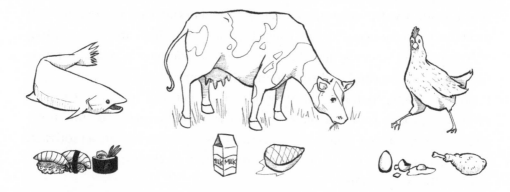

Animals as Food

One of the most visible uses of animals is as food. We are often so separated in modern society from the process by which living animals are converted into food that we can often forget (or ignore) its origins. Many people argue that meat-eating is "natural" and thus morally neutral because other animals eat animals as well. However, many vegetarians and vegans argue that we can have a healthy diet without eating animal tissue, and that it is unethical to eat animals.

While some countries such as India have strong cultural or religious traditions that promote vegetarianism, it is not a view shared by much of the world. In Western cultures, more than 90% of their populations tend to eat meat from animals, indicating that the use of animals for food is generally accepted by these societies.

Animals as Pets

Another highly visible use of animals is as companions, or pets. The most common pets are cats and dogs, followed by birds and fish. While some groups such as PETA advocate that pet ownership stems from our "selfish desire to possess animals and receive love from them" and that owning a pet "causes immeasurable suffering," this does not seem to be a widely adopted view. Even non-pet owners generally do not have a problem with dogs and cats, as they have been highly transformed from their ancestors (wolf and African wildcat).

Given the number of pet owners, it can be assumed that society is reasonably accepting of the use of animals as pets.

Animals for Work

Before the Industrial Revolution, which brought the invention of the steam and gasoline engines, much heavy work was done by animals, windmills, or waterwheels. Even today, horses and oxen are actively used on farms and ranches around the world. Dogs are used in security and law enforcement (guard dogs, K-9 units, and drug-sniffing dogs at border crossings). Dogs are of high assistance to visually impaired people, and some dogs can alert owners with epilepsy of oncoming seizures. Horses allow police officers a higher vantage point and maneuverability in crowds.

Some may argue that the animals are not truly "working," as they have had no choice. They were not "asked" if they would like to work, so, in reality, animals working is more closely related to slave labor. However, this view is not shared by a majority of the population, which generally agrees that if animals are reliably fed, sheltered, and cared for, then working can be a mutually beneficial arrangement for the animal and the human.

Animals for Research

Animals are also used extensively in biomedical research and investigations. One of the most famous research animals was Laika, the Soviet dog who became the first animal to orbit the Earth. These early space experiments using animals were conducted to test whether a living passenger could survive being launched into orbit and endure the microgravity environment. Dogs were also used in Pavlov's experiments on digestion, which led to the theory of classical conditioning. Pigs and other farm animals are often used as stand-ins for humans for testing medical devices. Monkeys are used when testing new vaccines and treatments for disease, as they more closely resemble us humans.

Society's view of the use of animals to study human physiology and ease human afflictions remains philosophically complicated. Some feel strongly that experiments on animals should never be done. Many feel that it should only be done when there are no other alternative models, and if the research stands to benefit human afflictions. Given the ethical complexities, all universities and research institutions have Institutional Animal Care and Use Committees (IACUC), where a board of experts review and approve, disapprove, or modify all proposed experiments involving the use of vertebrate animals.

Animals in Education

Finally, we discuss the ethics pertaining to the experiments in this book: the use of animals in education. Biology classes for the past 100 years have used preserved animals (frogs, fetal pigs) to help teach physiology and anatomy. College and high school students also work with living subjects for experiments which cause no harm to the animal. For example, rats are used to study learning and memory by solving mazes or mechanical puzzles to obtain food. Insects are used to teach about metamorphosis and used in this book to teach about the nervous system.

For the past few decades, there has been a movement to ban the use of animals in education. While some schools have stopped the practice altogether, many find that the use of animals has clear scientific and educational purposes in the classroom. Professional organizations such as the National

Science Teaching Association and the National Association of Biology Teachers support the use of animals in education, as interaction with organisms is one of the most effective methods of achieving educational goals in biology.

In this book, our general sentiment both ethically and legally is that students should not do any invasive experiments on vertebrates. We will limit ourselves to using invertebrates (animals without backbones, such as insects), as they generally have less complex nervous systems and are more robust to experimentation. In our invertebrate experiments, we try to develop "survival preparations," where the animal can continue to live a normal existence of eating and breeding after the experiment is completed. When designing our experiments on animals for education, we attempt to be as minimally invasive as possible, on the "simplest" animal possible (with the least amount of neurons), and with the maximum learning effect for the student.

An Ethical Framework for Using Animals

When determining whether it is ethical to use animals in a given situation, we feel it is important to think about a "cost-to-benefit ratio." Meaning, what is the cost to the animal versus what is the benefit to society. This isn't a numerical calculation, but rather a framework for philosophical thought. The decision point on when the cost exceeds the benefit will always be subjective, but it will allow for a thoughtful discussion of ethics.

In each of the uses of animals discussed above, it is possible to place the cost of the animal in general terms. For example, for food and often in scientific research, the cost to the animal is very high, meaning that it dies. For pets, the cost seems to be pretty low. Animals used in work could range from low to medium, depending on the task.

The benefit to society can also be placed on a scale. Humans need to eat, and animals provide a lot of calories; therefore, the benefit is high—though this may change over time. With efforts to make artificial meat, either using plants or by growing muscle tissue in labs, we may one day decide that there is little benefit from eating living animals over alternatives. The value to humans who own pets is high. In animal research, the human suffering alleviated from animal studies provides a large benefit to society.

Taken together, we can start to discuss what is ethical by comparing the two sides: costs and benefits. For example, pet ownership provides high value to humans with low cost to the animal. This suggests why owning a pet is not a very controversial animal use. Medical research provides many benefits, but at high costs to the animal. The ethical review boards at universities and research institutions look very carefully at each experimental procedure to determine whether they can be justified, or whether there is a way to replace, reduce, and refine animal experiments.

The Ethics of Using Animals in This Book

When developing this book, we considered this ethical framework. We have carefully evaluated the cost to the animal in our most common experimental procedure, removing the leg of a cockroach. We know that in the wild, the leg or antenna from an insect are often missing, and that they have evolved a way to grow them back. A cockroach's leg can easily detach, which is hypothesized to be a defense mechanism when grabbed by a predator. We have carefully looked at our classroom techniques and have documented the high survival rates and leg regrowth of the cockroach in a peer-reviewed experiment:

- Marzullo, T. C. "Leg Regrowth in *Blaberus discoidalis* (Discoid Cockroach) following Limb Autotomy versus Limb Severance and Relevance to Neurophysiology Experiments." *PLOS ONE*, 11, no. 1 (2016): e0146778. http://doi .org/10.1371/journal.pone.0146778.

So, the cost to the insect seems low given the survival and recovery rates, but what are the societal benefits of our educational neuroscience experiments? According to the World Health Organization, 20 percent of the world will be affected by a mental or neurological disorder at some point in their lives, and we have no known cures. What is required to advance understanding of neurological disorders is basic neuroscience research. Most people do not have even a basic understanding of how the brain works. Given the need for important neuroscience research in the future and the need to educate the public on neuroscience, we feel the benefit to society is high.

Given that the cost to the animal is low, while the benefit to society is high, we therefore conclude that the outlined experiments involving animals in this book are ethical. Others may have differing opinions on this matter, so it is always good to first have an ethical discussion before using animals in a classroom setting.

When performing the surgical techniques in this book, we make sure to always anesthetize the animals. We actually do not know if insects feel pain during procedures, but we make the assumption that they do and attempt to minimize their discomfort. Whether the insect feels pain when it wakes up from the surgery, we do not know. All we know is that the wound does heal, that the cockroaches are walking around within hours, eating lettuce, drinking water crystals, and making more cockroaches. They don't act or behave any differently than other cockroaches. We have established, and urge citizen scientists to adopt as well, a retirement community for roaches who have given a leg or antenna for science. These roaches can readily produce offspring for future experiments. See appendix 1 for a discussion of how to house and care for your cockroaches.

We close this discussion with a note to future historians that may happen to stumble upon this book while investigating animal ethics of the twenty-first century. At the time of this writing, our machine intelligence systems have not yet mastered biology. There is currently no replacement for experimentation; there is no model we can run to know how a complex biological or neural system will work. Perhaps that day will come, and we readily embrace its arrival.

PART I

NEURONS

I
Neurons

HEAR AND SEE A NEURON

2

Hear and See a Neuron

To understand a biological system as complex as the brain, it's good to start with the basic structural units of the brain: the cell. Cells form the functional building blocks which make up all living organisms. The brain contains many different types of cells that give it blood, protection, and structure, but we will be focusing on a cell with very unique properties: the *neuron*. These are the cells that can send messages from the eye that allow it to see and to our muscles to move. But how are these messages sent? How can these messages encode senses or make movements? We will start off with an experiment.

Our brain is made up of about 80 billion neurons (that's 80,000,000,000), all sending messages and working together to form our consciousness. And while it is possible to record these signals in humans, you would have to stick sharp needles inside your brain or nerves to gain access to neurons. However, we can take advantage of evolution to learn about ourselves from a more practical stand-in. About 500 million years ago, the nervous system of our great ancestors (back when we were still fish) partitioned into a brain and a spinal cord–type structure. These ancient nervous systems are well conserved through evolution, meaning that the animals derived from those early days all tend to have very similar neurons. All the animals on earth have neurons that are roughly the size of ours and act in ways that are similar to ours. It's not the size of the neuron that makes us different—it's the number of our neurons. In this book, we will take advantage of this fact by looking at a wide variety of animals. Many can be found in your own backyard, or in pet stores as feeder animals.

In our first experiment, we will be using a cockroach (*Blaberus discoidalis*). Cockroaches have about 1 million neurons (80,000 times fewer than we do), but these neurons give rise to some similar behaviors, such as sensing, eating, and running. By understanding a cockroach's nervous system, we are really beginning to understand our own!

CENTRAL NERVOUS SYSTEM

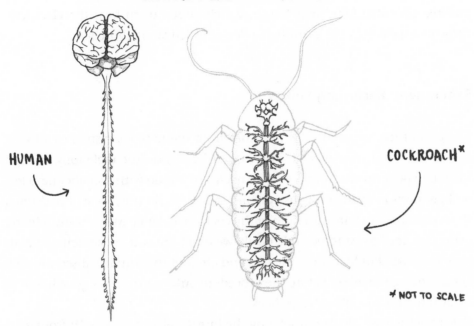

HUMAN

COCKROACH*

* NOT TO SCALE

So let's get started by trying to detect these hidden messages being sent by neurons! Ever since Luigi Galvani's work in the 1780s, scientists have known that there was a relationship between electricity and muscle movement (more on that later). Given that electricity seems important, we will first attempt to detect any electrical activity from within the cockroach. To do this, we will use a scientific instrument called a bioamplifier.

Bioamplifiers allow us to measure electricity in living tissues. Like many things in science, we need to make things bigger to study them, and that is exactly what a bioamplifier does. Telescopes make distant planets big enough so we can observe them, microscopes allow us to see small cells, and PCR machines amplify segments of DNA so we can detect them. A bioamplifier takes small currents of electricity and makes them bigger so we can see and hear them. We will be using an open-source DIY bioamplifier called the SpikerBox in these experiments. You can order one online, or if you prefer to build one yourself, you can find instructions in appendix 2. The SpikerBox electrode has two metal needles that allow electricity to flow into a series of amplifiers, making it about 1,000 times bigger, which allows us to hear the electricity through a speaker or see it through a device called an oscilloscope.

Experiment: Recording Spikes

The goal of this experiment is to record from the nervous system of a living cockroach. Since your average roach won't lie still on the table long enough for you to perform an experiment, we need to make do with a smaller (and less mobile) portion of the nervous system: the leg. To do this, we will perform a short procedure to remove a limb of the cockroach. Since we are using a living being, we need to be respectful. While we don't know if cockroaches experience pain, we should assume they do and attempt to minimize discomfort by anesthetizing the insect before the procedure. Afterward, it will get a home to recover within and grow the leg back.

Let's get started with anesthetizing the insect. Select a cockroach from your container (gloves recommended!) and put it in a glass of ice water. Cockroaches are cold-blooded, so its body will quickly acquire the temperature of the surrounding water. As the cockroach cools, its internal systems also slow down and will slowly become anesthetized. Leave the insect in the ice water for a few minutes until it stops moving. The cockroach will not drown during this process, due to a lowered metabolic rate and surface tension that does not allow water to rush into the body's breathing pores (called *spiracles*).

Remove the anesthetized cockroach from the glass and place it on its back on a flat surface. Now it's time for the surgery. Gently pull off one of its legs, gripping it near the body. Don't worry, the leg is designed to break easily at this joint (like the tail of a lizard) and will grow back to full size within 125 days. You can also cut the leg off using scissors, but we have discovered that pulling the leg makes the cockroach regenerate the leg faster.

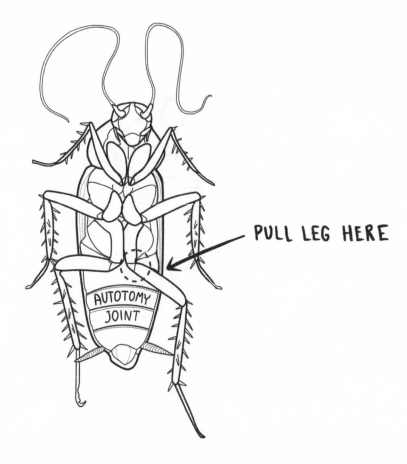

Return the cockroach to its home container to recover. It will start waking up and slowly start moving in a few minutes. About an hour later, it will be scurrying away and behaving like normal.

Place the leg you removed onto a piece of cork or balsa wood.

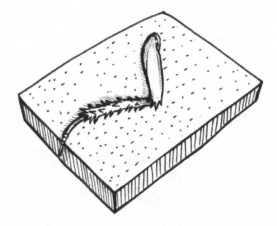

Insert the two electrode pins in the leg and make sure they are pushed into the cork/balsa wood backing. The placement doesn't really matter at this point, so just spread the leg out and place the pins through the center of the leg at two positions. Once in place, you can connect the wires to the SpikerBox.

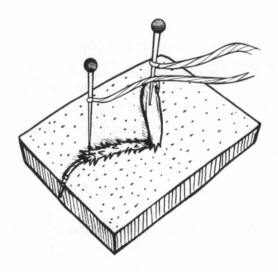

Turn on your SpikerBox and listen to the amplified electrical output as it passes through the speaker. Listen carefully. You should begin to hear some patterns in the noise: popcorn-like sounds, or perhaps the sound of rain falling on a roof. Congrats! You are now listening to the electrical messages of the nervous system! Now let's try to see what these electrical discharges look like.

If your smartphone or computer has a microphone port, then you can easily visualize the electrical signal using our open-source data visualizer, SpikeRecorder. Connect the output of the SpikerBox to your smartphone or computer.

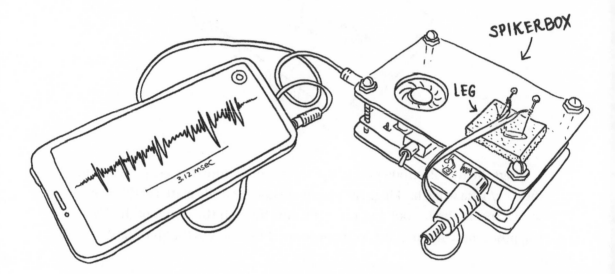

On the screen, we can now see what that popping sound looks like. There are little peaks of electricity (see arrows). These are known as "spikes," and they are what the nervous system uses to convey messages. You may notice that there are various sizes and shapes of these spikes. Now, we can't read out what the spikes mean at this point . . . we are just eavesdropping in on a random conversation between neurons.

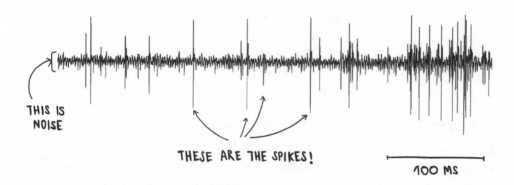

Let's explore a little further. The spikes are passing by fairly quickly, far too fast for us to get a good look at them. So, we will use a technique to stop the action during the spike, which will allow us to see what is going on. To do this, find and select the "threshold" view on SpikeRecorder. This sets up a trigger that will pause the spike every time the voltage passes a certain threshold. If you set the threshold around the peak of the spikes, you can zoom in and see an individual spike in all its glory.

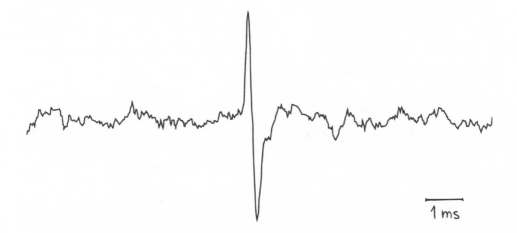

1 ms

This is the spike: the key source of electrophysiology. The spike is the basic information-processing unit of the nervous system, through which all neurons communicate. It is the common currency of the brain, the euro of neurons, in which all our senses are converted, all our thoughts are processed, and all our muscles move.

The Spike

Given the importance of spikes, it would be good to understand a bit more about them. We can start first with the name. A "spike" is a colloquial name that neuroscientists use for an "action potential." You may hear a scientist use the words "spike," "action potential," or "electrical impulse" interchangeably while discussing the phenomenon we just observed, but "spike" is the simplest, so we will stick with it.

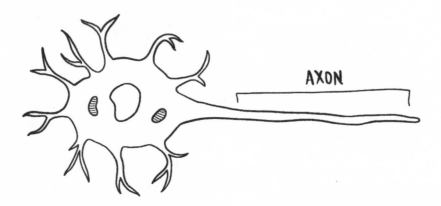

Spikes are transmitted by neurons, the cells of the brain. Let's take a look at a neuron and what makes it different from other cells.

On the surface, a neuron is like any other cell of the body, with a nucleus containing DNA and other cell structures. What starts to make it look unique are the structures that pull away from the cell body, which are called "processes." There is one particular process, called the axon, that is considered the "output" of the neuron. It is down this axon that the electrical spikes travel from the cell body outward.

One question you may ask is: Why do neurons use electricity? Using electricity may seem strange given that many biological systems such as bacteria use chemicals to communicate. While that method works well, it is limited by the speed at which chemicals can diffuse.

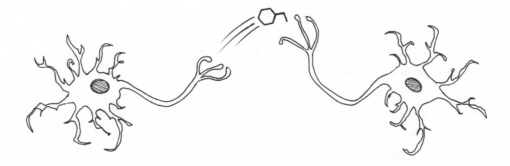

Chemical diffusion is slow over long distances. Think about how long it takes an odor to travel from one side of the room to another. When you chop up an onion, how long does it take for someone sitting across the room to smell it or start crying? It takes time! This speed wouldn't work well in the brain. If you were about to get hit by a bus on the street, you wouldn't want your eye to diffuse a chemical message that tells you to jump out of the way. You would want a lightning-fast signal that makes you act in a split second. This is where electricity comes in: your nervous system uses electricity because it is much faster than chemicals.

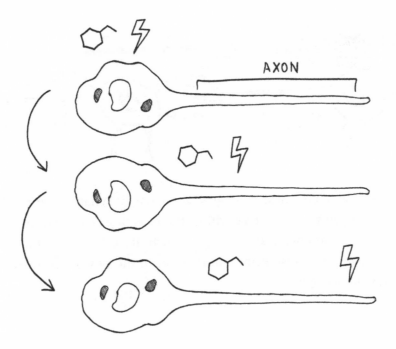

But that doesn't mean that our brain is entirely electrical! If electricity were the only signal used in the brain, things could spiral out of control fairly quickly. There would be no balance. All the positive electrical currents from spikes would accumulate, and soon cells would all reach their maximums. To avoid this, neurons almost always use chemicals to communicate with each other. Chemicals allow the cells to either increase or decrease their voltage potentials (depending on the type of chemical received). But as we noted before, chemical diffusion is slow! To speed things up, neurons press themselves incredibly close together so that the chemical messages only have to travel short distances (20 nm or 0.00000002 m). This space where two neurons meet is called a synapse. Chemical diffusion has a strange property that makes it happen quickly over short distances, but slowly over larger ones. For example, if a chemical message, say, glutamate, had to travel down an axon, it would take over 16 minutes just to travel 1 mm. But diffusing across even a large (50 nm) synapse to your muscles can happen in only 2.5 us (0.0000025s). In fact, the gap between neurons is so small that even though it was hypothesized to exist in the early 1900s, it wasn't until the scanning electron microscope was used in the 1950s that scientists were actually able to see it.

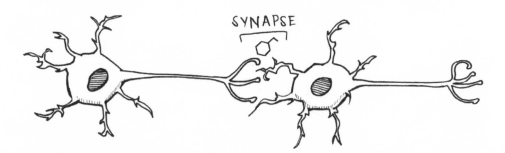

SYNAPSE

It is the type of chemical message sent across the synapse that determines what will happen to the voltage within the receiving cell. Neurons only use a handful of these chemical messages, which are called neurotransmitters. These messages are transmitting information to the next cell on whether to gain or lose its voltage.

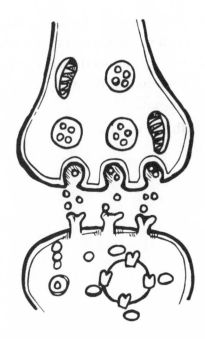

When a spike travels down the axon and reaches a synapse, the electricity causes a small amount of neurotransmitters to be released onto its neighboring cell, slightly changing the voltage. If the voltage gets high enough, it will cause a spike to travel down down its axon, and the cycle continues on to the next neuron.

So that's what a spike is . . . but what does a spike mean? And how can a single spike carry so much information? We will continue to use our cockroach leg to begin to answer these questions in the following chapters.

Follow-Up Questions

(1) What would happen if you attempted this experiment on another insect—say, a grasshopper or beetle?

(2) Does the cockroach leg have to be "alive" to record spikes? What happens if you record from the leg a few days after the surgery?

(3) We saw that the ice water knocked out the cockroach. Is this because of its nervous system? What would happen to the spikes if you were to cool down the cockroach leg? Try putting the leg in a cold refrigerator for a few minutes, and then record from it. What happens? How does this relate to anesthesiology?

(4) Why even bother with the cockroach? Why can we not record directly from human neurons?

NEURONS FOR TOUCH

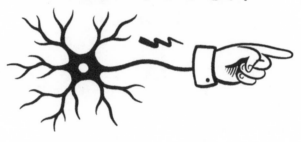

3
Neurons for Touch

Touch sensations are very important to both us humans and cockroaches. We use tactile feedback from our fingers to guide us while we eat, draw, write, and type. Cockroaches, too, are highly sensitive to tactile inputs. They react very quickly to subtle changes in the room around them to evade their predators, including us humans. They are always on the lookout for danger, such as approaching footsteps or a door swinging open, and they react very quickly.

One reason people find cockroaches creepy is that they are one of the fastest insects in the world! They can cover 15 body lengths per second (as fast as a cheetah), which would be like a human running at 62 mph (100 km/h)! By the time you turn on the lights in your kitchen at night, the cockroaches have already sensed you coming and are at full speed heading to safety. But how exactly does the cockroach's brain "know" when there is an unexpected breeze in the air or a vibration in the ground? Let's look a bit deeper at the sensation of touch.

Experiment: Somatosensory Neurons

The goal of this experiment is to gain an understanding of how neurons in the cockroach leg encode tactile information. We will be using our cockroach-leg preparation method from the previous chapter, only now we will examine the responses a little more carefully. For this experiment, let's place the two pins a little closer together in the leg.

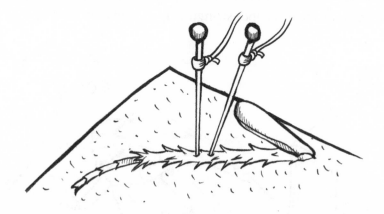

The closer you move the pins, the smaller our recording area becomes. For this experiment, we want to record from a small area so we see spikes from fewer neurons. But not too close! It is important that the metal pins don't touch each other, or else they will short out and not be able to record. With your pins in place, turn on the SpikerBox and start listening to the electrical activity. It could be rather silent, or you may hear some clear random pops from spikes. But let's see what happens when we puff a light wind by blowing on the leg.

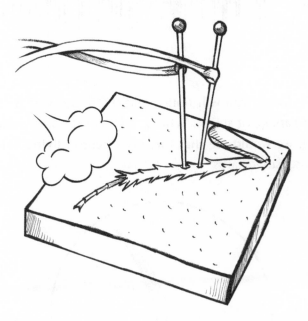

You should notice that each time you blow, the robust sound gets louder. Open your SpikeRecorder app again, and let's take a look at what is happening. Zoom out a bit so we can see about 10 seconds of time. Blow on the leg a few more times and observe any effects.

If you look closely, you will notice that the sound you are hearing when you blow is a barrage of spikes. What is happening? If we were able to look inside the leg, we would see that there is a neuron living within the cockroach's touch-sensitive barbs.

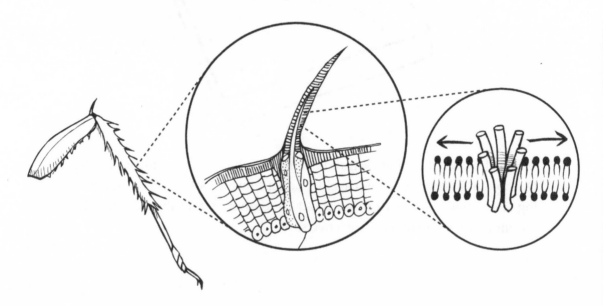

As the wind presses against the barb, it slightly stretches open small channels on the surface of the neuron, which allows some positively charged sodium ions (Na+) to rush in. This changes the voltage in the cell, and when it reaches its threshold, it fires a spike. The axon carrying the spiking activity eventually synapses on a neuron that would travel to the brain, alerting the cockroach that something is afoot.

At this point, you might be wondering . . . How is this neuron still functioning? After all, it was detached from the body! Shouldn't the cells in the leg die? Why would it still send spikes? The answer to these questions is that the neuron thinks everything is fine. Cockroaches breathe through small spiracles in their skin. When the leg is removed, a small hole is formed that still allows oxygen and carbon dioxide to diffuse easily. So the neuron can carry on doing its job, sending messages when it detects vibrations. It will eventually dry out, but it has enough energy reserves to last for days. This property makes this cockroach-leg preparation very useful for long investigations. As pin positions dry out, you can just move them to a new location to continue.

Experiment: Somatotopy

Back to the experiment! Let's explore this train of spikes a bit more closely. Instead of blowing on the leg, we will be a little bit more careful with how we stimulate the touch neurons.

Again, we want to keep the electrode pins close (but not touching) so that there is almost no spiking activity happening spontaneously. Next, take a toothpick and gently touch the barbs of the leg. You may notice that touching most of the hairs will have no effect on the recording, but you should eventually be able to find a hair that causes some spiking activity. Notice how quickly the sound of the spikes hits your ear as you touch the leg. Seems instantaneous! Neurons can respond and fire quickly. If you keep exploring more barbs, you may notice that different barbs show different trains of different-sized spikes.

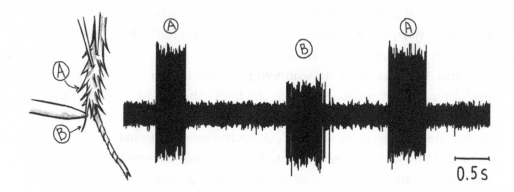

You can go back and forth and touch two different hairs to get two unique spike shapes. You will notice that the spikes from the same hair always tend to be the same height and shape. This highlights an interesting fact about spikes: an action potential almost never changes its shape. It repeats a spike over and over again in near-exact copies. So if you see spikes of different sizes or different shapes in your recordings, that's a good clue that you are detecting spikes from multiple neurons.

We are also beginning to see how touches on different parts of the body can be differentiated from each other. A neuron in the cockroach brain receiving the messages shown here is easily able to tell the difference between the two hairs. We call this correspondence between the receptor location on the body and the neurons in the brain "somatotopy."

The spatial arrangement of your body has strikingly similar organization in your brain. For example, you have touch receptors in your fingers and palm that synapse with neurons that go to specific parts of the spinal cord.

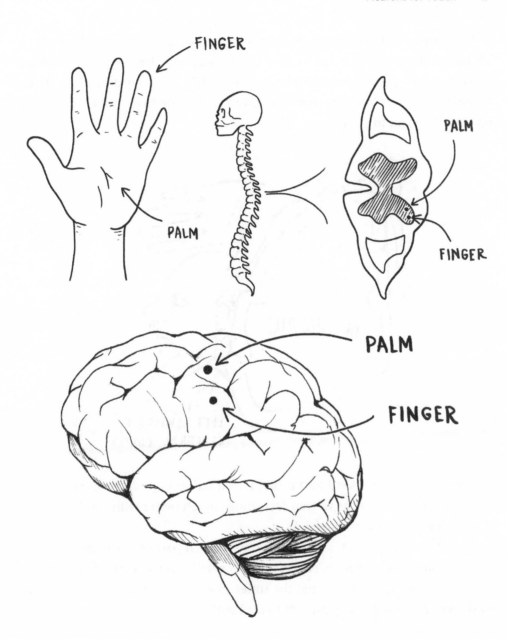

Those neurons then synapse with specific parts of the neocortex (the wrinkly part of the brain) in an area called the primary somatosensory (S1) cortex. The cells in S1 cover your entire body but are laid out a bit differently than you might expect. Neurons that respond to toes come first, and then follow up your body out to your arms.

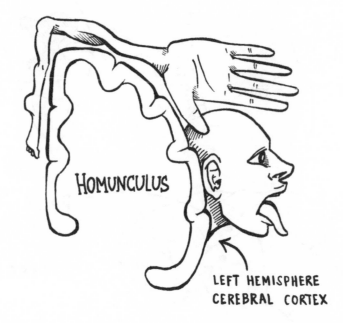

Neurons that correspond to our hands and face occupy a lot of real estate here. This makes sense, as we can feel things much better with our hands or tongue than we can with our backs.

So we now understand how we can sense which part of our body is being touched. But if the spikes never get bigger or smaller (always staying the same shape and size), how can we tell the difference between a light touch and a hard touch? Let's do an experiment to find out!

Experiment: Rate Coding

For this experiment, we are going to change how much pressure we place on the leg hairs. Find a hair that elicits a large response when you press it with a toothpick. Once you find a suitable candidate, practice pressing on it lightly and then a bit harder. Press the barb for only about half a second, then release it. Try little pokes of various pressures. You will notice that the hair will bend a bit more the harder it is pressed. Now let's look at the train of spikes for three different pressure types: light, medium, and hard.

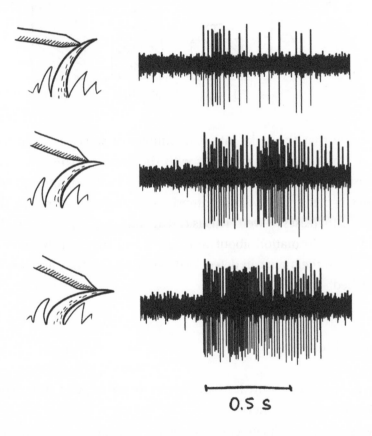

0.5 s

    The results here look rather interesting. Even though you press harder on the leg, the spikes stay the same size! However, there are more spikes. We can quantify this by counting the number of spikes that occur within a fixed amount of time. In SpikeRecorder, there is a feature to "find spikes," which makes counting a bit easier. But you can do it by hand too. Let's count the spikes in the first half of a second. We can then plot the results.

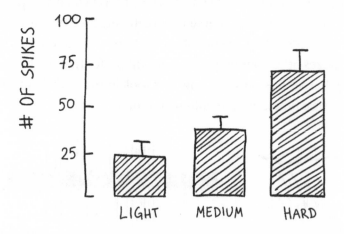

We can see that the spike rate (the number of spikes released in a fixed amount of time) definitely increases with the pressure. This means that this touch-sensitive neuron alerts the brain to being touched, and indicates how much force is being used. As the intensity of a stimulus increases, the firing rate of spikes increases as well. This is called "rate coding," as the spiking rate is "encoding" information about touches. Taken together, we now can see how neurons in the brain can detect both *where* on the body and *how hard* it is being touched!

Follow-Up Questions

(1) Why do you think some leg barbs are more sensitive to stimulation with the toothpick than others?

(2) If you are in a school lab, you might have access to an air cylinder, which allows you to control the amount of pressure. Set the air tube at a fixed distance away from the leg, and apply increasing pressure from 0 to 30 PSI

in small increments. Can you develop an equation for how the neurons encode air pressure?

(3) It would be better if you could quantify the force used when touching the cockroach barb rather than pressing "light" or "heavy." Fishing lines have an interesting property where if you press a piece of it against an object, the force applied is constant as the line bends. The longer the length of the fishing wire, the lower the force. You could glue different lengths of fishing line onto popsicle sticks, and determine their force by pressing them down on a letter scale and noting down the weight. Can you quantify this experiment more carefully? Can you develop an equation (spikes per gram) for the leg hairs? Are the rate coding equations the same for different barbs up and down the leg?

HOW FAST ARE NEURONS?

4

How Fast Are Neurons?

In our cockroach-leg experiments, we saw that neurons can take the information generated by touch, encode it in the number of spikes per second, and send those spikes along to the brain for further processing. We will see that this is a general way in which our senses are represented in spikes. But what was missing from this experiment is speed. How fast can those spikes travel down the axon to reach the next synapse?

The nervous system is pretty fast. In our own bodies, we can barely tell the difference between wanting to move our hand and actually moving it. In our cockroach-leg experiment, we seem to hear the spikes immediately when we touch the leg of the cockroach. But it is not instantaneous. Not even light, the fastest signal in the universe, travels instantaneously. It takes time for the signal to transfer from one area to another. But how fast is the nervous system? Is it faster than a bicycle, a plane, or as fast as the electricity in your house? How can we measure it?

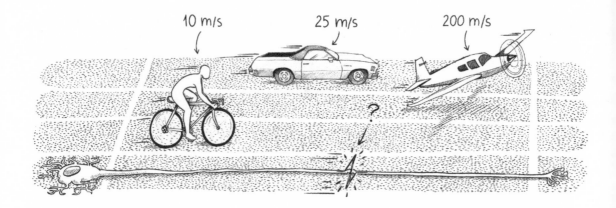

So far we've only recorded our neurons using one channel, meaning we used only one recording electrode and one ground to gather data. We could measure the time it takes for the voltage to go up and down to form the spike, but not how fast the peak of that spike travels down the axon. To measure speed (velocity), though, you need to measure both time (when a spike occurred) and distance (how far a spike traveled down a nerve).

As an analogy, think of a car driving on a highway. If you were looking out of a small observation hut by the road, you could tell whether you saw a car pass you, what kind of car it was, and the moment when you saw it go by.

Suppose you had a friend 1/2 mile down the road in a similar hut, and you each wrote down when you saw a specific car pass your hut. Later, you two could compare notes, and with some quick math, you could determine the speed of the car. Say it took one minute for the car to pass between the two observation huts. Convert that minute into hours (1 minute = 0.016 hours), and dividing the distance (1/2 mile) by that, you would calculate a speed of approximately 30 mph.

Speed is a function of distance over time, so we can only quantify it with at least two observers. For the same reason, we need two observers (two recording pins) to measure at two points along a nerve as a spike travels down it. If you only have one channel, you can tell if you saw a spike and when it arrived. However, you couldn't tell how fast the spike was traveling down the axon.

Luckily, our cockroach leg is large enough to support two recording channels, allowing for two different observers! To measure the speed of spikes, we need to place two separate recording pins in the same leg, measure the distance between two pins, and get ready to note down the time the spikes arrive at both the first pin and the second. But when we set this up, you will notice a problem immediately. There are a lot of spikes happening on both channels. In fact, there are way too many spikes to keep track of which spike on the second channel was the same spike that passed by the first. They all look the same!

This problem could happen on the road, too, if we go back to our cars visualization. Imagine a very busy, fast-moving freeway in Detroit with many similar-looking cars. You can see the problem!

It would be hard to determine which car you and your friend were trying to time if there were multiple cars on the road of the same make and model.

Similarly, there are a lot of spikes occurring in the cockroach leg, and identifying unique ones with just two observers is very tricky. The femur of the cockroach leg has 200–400 axons, all firing many spikes. We also don't have a lot of space to place the electrodes, since cockroach legs are only about 8 mm long. So this preparation is not very conducive to our speed experiment. We shall have to contrive another!

Fortunately, we don't have to look far in the wonderful world of invertebrates. Our goal is to find an animal that we can record from only a handful of long axons, and whose neurons do not fire many spikes. Could there be a creature in the animal kingdom that meets all these criteria? The answer is yes! It is probably living right under your feet as you read this: the common earthworm!

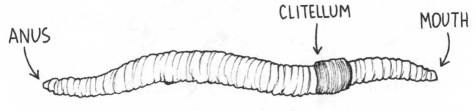

ANUS   CLITELLUM   MOUTH

*TOP (DORSAL) SIDE OF WORM DARKER THAN BOTTOM (VENTRAL)

We began our journey into neuroscience using an insect (or arthropod), but now we are adding a new class of invertebrates to our roster: worms (or annelids)! The common earthworm, or *Lumbricus terrestris*, has far fewer neurons than the cockroach, and its body is much longer. To make it even better, the earthworm contains only three large axons that run its length: one medial giant fiber and two lateral giant fibers. The medial giant fiber transmits information about the front of the worm, the part closest to the clitellum, and the lateral giant fibers transmit information from the skin cells of the back end of the worm.

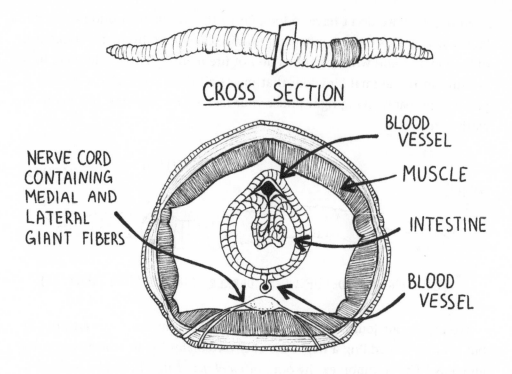

NERVE CORD
CONTAINING
MEDIAL AND
LATERAL
GIANT FIBERS

CROSS SECTION

BLOOD VESSEL

MUSCLE

INTESTINE

BLOOD VESSEL

Experiment: Earthworm Conduction Velocity

The goal of this experiment is to measure how fast spikes travel in earthworms. We call this measurement the conduction velocity of spikes. To get started, you will need to purchase some worms! Common earthworms are used for fishing and can be found at your nearest pet store, sporting goods store, or gas station. You can even dig them up if you live in northern wet climates where earthworms can be found.

Next, you will need to anesthetize your worm. Our ice water bath will not work on earthworms. They are used to cold weather and have evolved neurons that work at low temperatures. So we will need to prepare a 10% ethanol solution. If you are not in a school lab with an ethanol stock, there is an easy way to do this at home. You can use vodka (which is normally 80 proof, or

40% ethanol) and dilute it further using 1 part vodka, 3 parts water. For example, we mix 10 milliliters of vodka with 30 milliliters of tap water.

Pick out an earthworm from the soil. If you scoop up a worm that is not wriggling and squirming in your hand, then it may not be healthy and you may not get good recordings. Make sure you have a healthy worm and place it in the alcohol mixture for 3–4 minutes. Do not wait too long; as with human anesthesia, the delicate balance between too little anesthesia and too much is tricky. Too little anesthesia, and the earthworm will move around during the experiment, resulting in a lot of movement noise. Too much anesthesia, and the nerves will not fire. About 3–4 minutes seems to be a good range.

Place the earthworm on your recording base: balsa wood, cork, or styrofoam will work. Insert three pin electrodes from your SpikerBox along the worm. Spread out the two recording pins to have roughly the same distance between them, and arrange them so they are channel 1, channel 2, then the ground pin. Push the pins through the worm, but slightly off center (so you don't damage the nerve fibers), and into the platform underneath. Once everything is in place, be sure to measure the distance between our two recording pins (channels 1 and 2) and write it down. It's always good to take a picture of your setup for reference.

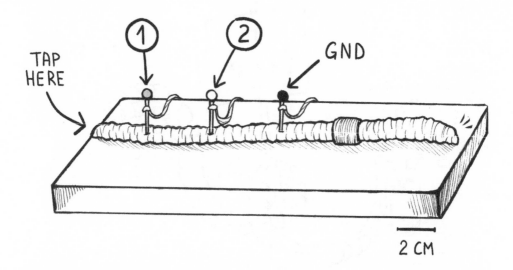

Open up your SpikeRecorder app and press the record button when you are ready. Now for the stimulus. Using a small plastic probe like a coffee stirrer, tap the posterior (back) end of the worm. You should hear the evoked spikes caused by the tap. Aha! Spikes! Tap the end of the worm 3–4 more times, waiting about 3–4 seconds between each tap so that they are distinct on your recording.

Once you have several spikes, you can stop recording and remove the electrodes from the worm. Dip the worm briefly in water to remoisturize it, and return the worm to the soil in its box where it can sober up. The earthworm is quite resilient and recovers well from this experiment. It can tolerate the needle placement and can be used for another experiment on another day. If you dug it up, you can return it to the environment where you found it.

It's time to see how fast these spikes were traveling! Open up the software again and look at the recordings from both channels. You should see the spikes you heard while recording for each tap on each channel. They come in clusters of 1–3 spikes.

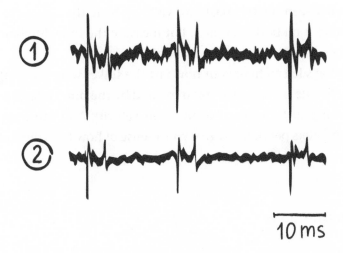

10 ms

They look like they are happening at the same time, but if you zoom in more, you will notice that the spike first appears on channel 1, then a bit later on channel 2.

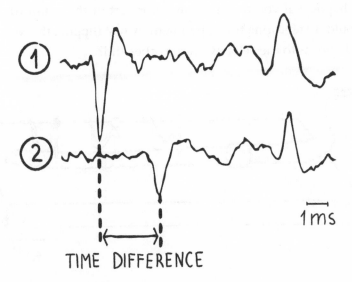

TIME DIFFERENCE

1ms

The spike produced by the tap had to travel down the worm and pass by both electrodes! Measure the time delay between the two channels by looking at the time between the two peaks (or valleys). Note that the spikes may look slightly different on both channels, but measure the time difference between the first large deflections on both channels.

Now all that's left to figure out how fast the spikes were traveling is to do a little math. Divide the distance you measured by the time difference recorded. Voilà! You have just measured conduction velocity. You can convert this to miles or kilometers per hour, to give you a sense of how fast your brain is sending thoughts.

Follow-Up Questions

(1) You tapped the worm a few times. Does the speed measurement change from tap to tap?

(2) Does the speed measurement change from earthworm to earthworm? Are smaller earthworms faster or slower than large earthworms?

(3) What happens if you touch the anterior part of the worm (the mouth)? You could try repeating this experiment by first flipping the worm around. What happens to the velocity on the other end?

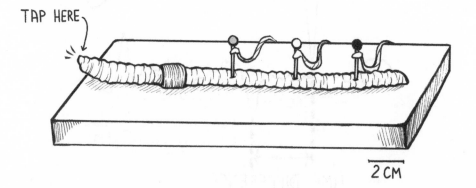

NEURON STIMULATION

5

Neuron Stimulation

So far we've been able to see and hear the electricity produced by neurons. Information is encoded in electrical spikes sent quickly down to the end of the axon. We have been able to "read" from the neurons using small metal electrodes; might it also be possible to "write" to them as well? Could we reverse the electrode and send electricity into a neuron?

Experiment: DC Microstimulation

Our goal is to test if we can communicate with neurons using electricity. Let's start with our cockroach-leg preparation. For this experiment, we found that it works best to place the pins in the femur and coxa.

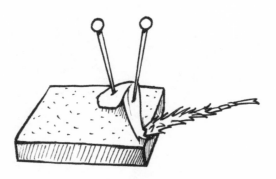

Tape two wires on a 9V battery, one to the positive post, and the other to the negative post. Touch one wire to each of the pins so that the electrical circuit is completed within the cockroach leg. What do you see?

It's alive and (almost) kicking! The electrical current appears to be interacting with the neurons, causing movement! What could be happening? The answer is deeply rooted in the history of neuroscience.

Long before scientists were able to record spikes, they were able to see the relationship between the nervous system and electricity using simple batteries. In 1780, an Italian scientist named Luigi Galvani made a remarkable discovery: when electricity was applied to the nerves of frog legs, it caused the large muscles to twitch.

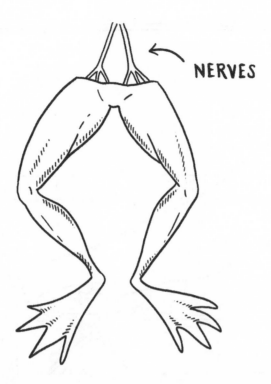

This discovery led to an interesting scientific debate as to whether "animal electricity" was the same as the electricity seen in the world, for example during lightning storms. Galvani tested this by hanging frog legs off his back porch during thunderstorms and watching the legs twitch.

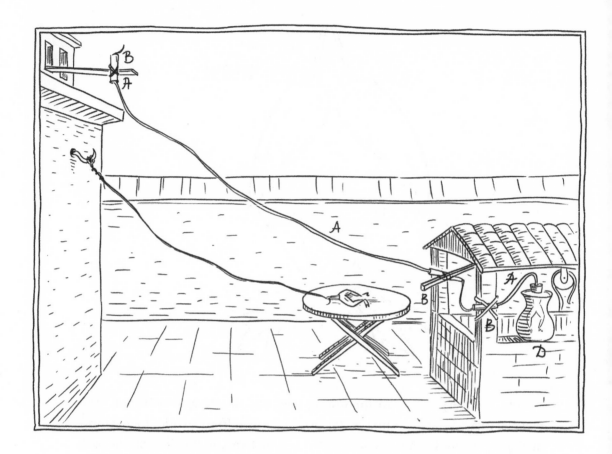

These phenomena became known as "Galvanism," and it made a lasting impression on people. There were fears that scientists could use electricity to restart life! Electricity itself was a relatively new and poorly understood force, so it seemed plausible that it could be used to make creatures come alive after death. In fact, Mary Shelley even claimed these experiments as a direct inspiration for her famous 1818 novel, *Frankenstein*.

> Perhaps a corpse would be re-animated; galvanism had given token of such things: perhaps the component parts of a creature might be manufactured, brought together, and endured with vital warmth. (Mary Shelley, introduction to *Frankenstein*)

Today, neuroscientists and neuroengineers use this electrical stimulation phenomenon to communicate directly with the brain. With a technique called "microstimulation," a small amount of electrical current (around 1–10 mA) is used to communicate with a small patch of the electrically excitable cells. In the cockroach-leg experiment, we were using the 9V battery to microstimulate the muscles directly and then observe the twitches.

Microstimulation in modern medical devices does not use direct currents as seen in Galvani's experiments or in our 9V battery. In practice, it has been found to be more efficient (and better for the neurons) to use pulses of both positive and negative currents.

Experiment: AC Microstimulation

We want to test if alternating currents can stimulate the nervous system. We will need an electrical device that can produce a moving current using small wires we can connect to our cockroach leg. Well, you are in luck, as you have been carrying such a device in your pocket! The headphones of your mobile device use alternating electrical currents to vibrate a cone to produce sounds and music in your headphones.

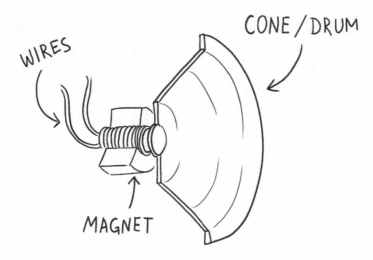

If you have a pair of headphones to sacrifice, you can cut off one of the earbuds and connect its two wires directly to the two pins in the leg. This will create a direct connection between the phone's output current and the cockroach's nervous system.

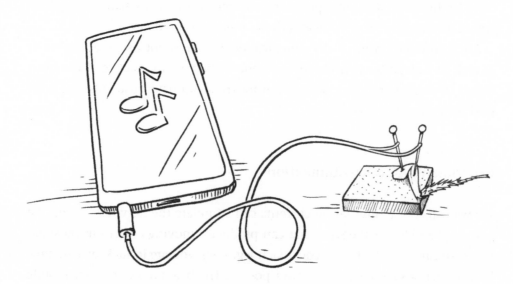

Now we are ready for the experiment. Load up your favorite music app, and choose two songs for your experiment. To illustrate the differences, one song should be in the hip-hop genre (our roaches prefer anything from the Beastie Boys album *Paul's Boutique*) and another in classical music (such as the *Goldberg Variations* by J. S. Bach). Start with a low volume (meaning low current delivered to the pins) and slowly start to increase it. Note down the song and at what volume you began to observe changes in the leg. Repeat the process a few times for each song. Which of the two songs made the legs move more easily?

So why exactly does your roach seem indifferent toward poor Johann Sebastian? It's not personal. When it comes to microstimulation, you gotta have bass! Club music, hip hop, and R&B music have steady bass lines that drive the rhythm. These bass frequencies are low, slow, and big!

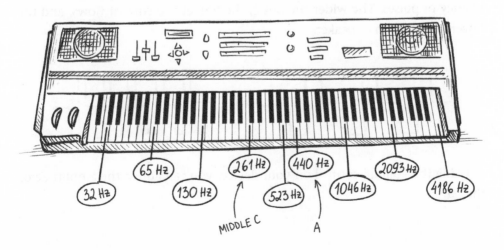

Chamber music on the other hand is mostly filled with instruments in the treble clef. The notes are higher, and the sounds much higher in frequency. So why does the frequency matter?

Recall that when the voltage changes in the cell, tiny channels physically open to allow the ion currents to flow in and out to produce a spike. This opening is fast, but does take a bit of time. If we change the voltage slowly, as in with low bass frequencies, these channels have time to open, allow ions to pass through, and initiate a spike to travel down to the muscle. But if the voltage changes too quickly (as it does for treble clef frequencies), the channels will begin to open, but close before ions can pass through and generate a spike. To find the "sweet spot," we can do one more experiment.

Experiment: Frequency Analysis of Microstimulation

In this experiment, we want to determine what frequencies can cause neurons to fire the easiest (meaning with the least amount of current). The lower the current, the more optimal the frequency is for exciting the nervous tissue. Biomedical engineers measure this when deciding how to stimulate the brain.

The setup will be the same as our music stimulation. Only this time, instead of playing music into the cockroach leg, we will play just tones. You will need to download a tone-generator app for your device. There are a lot of free ones out there; any will work. When you change a tone, you are changing the frequency of pulses. The wider the pulse, the longer the current flows, and the lower it sounds on a speaker.

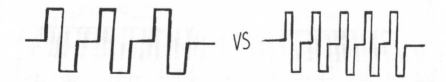

Using the volume control on your device, you can adjust the amplitude of the pulses.

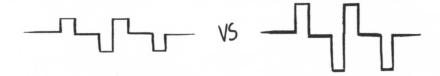

The higher the amplitude, the more the current is delivered, and the louder the sound. So our task is to determine which tone (pulse width) can deliver spikes with the lowest volume (current). To do this, we can make a chart with rows for four volumes (25%, 50%, 75%, and 100%) and with columns for frequencies (Hz) ranging from bass to treble (20, 50, 100, 200, 500, 1000, 2000, 5000).

	¼ VOLUME	½ VOLUME	¾ VOLUME	FULL VOLUME
20 Hz				
50 Hz				
100 Hz				
200 Hz				
500 Hz				
1 KHz				
2 KHz				
5 KHz				

⊕ MOVEMENT

⊖ NO MOVEMENT

Choose random points on the table and play the designated tone at the designated volume for one second. Write in the table a "+" if the leg moves, and a "–" if it doesn't. Run through the table a few times to get a good estimate of each pair. It's important to randomize, as the leg can get tired out from kicking around. Randomizing will help eliminate biasing the dataset. When you are done, take a look at the probability of getting movements ("+") in each square.

When you are all done, you can analyze the results. Visualizing the +'s can be hard, so you can plot the probability by taking the number of + trials (movement) divided by the total number of trials. This makes it a bit easier to see a pattern.

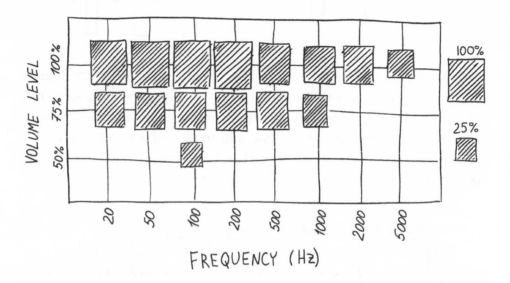

Here we can see that microstimulation at 100 Hz allowed neurons to fire at the lowest current setting. The bass range is between 20 and 200 Hz, so this result is consistent with our previous finding that the leg responded more to music containing a lot of bass!

Once again, the cockroach neurons are not that much different from our own. There are a number of clinical and research devices for humans that use microstimulation at roughly this same frequency for this very same reason. For example, patients with Parkinson's disease can be treated using deep brain stimulation (DBS). By inserting a small, long electrode into a specific part of the brain called the subthalamic nucleus, the shaking and tremors associated with the disease can be lessened. The stimulation frequencies used in DBS are typically around 100 Hz.

Follow-Up Questions

(1) What kinds of music cause what kinds of responses from the leg? Try a few different genres. Why do you think certain frequencies generate different degrees of response?

(2) How long can the leg maintain a constant flex from constant stimulation? Does the level of stimulation change this time? What causes it to stop? How long after removal before the leg stops responding to music? If it stops moving to music, does it still produce spikes on a SpikerBox? Why might that be or not be the case?

(3) Does the placement of the electrodes affect the response you get?

NEURONS FOR SMELL

6

Neurons for Smell

Each of our senses have developed their own unique way of detecting and encoding their stimuli. We've seen that the touch receptors open when stretched by physical contact, but what about smell? Our next task is to explore the peculiar dance of chemistry and electricity that allows our brains and bodies to receive odor information from the world around us.

The sense of smell is rather strange. Next time you smell rotting garbage in an alley, tacos from a food truck, or the fragrance of a dogwood tree, take a moment to think about the complexity of this sensation. An object in the world around you is sending you a coded signal by releasing small chemicals into the air. Recent studies have shown that humans can distinguish more than 1 trillion different odors!

The brain interprets these odors based on past experiences or genetics. Rotting garbage shrieks "Danger! Do not eat," which we interpret as disgusting. The smell of grilled food signals a high-calorie meal and is perceived by your brain as pleasant! We are not alone. In nature, there are countless examples of how smell plays a key role in helping animals determine what is good, bad, dangerous, edible, sexy, and more!

✷ SIZE OF MOLECULES NOT DRAWN TO SCALE

To investigate how this system works, we will turn to a new model organism that has shown a remarkability to draw on the sense of smell, the silkworm moth. The story of the silkworm moth is a strange one. If it turned into a movie, we humans would definitely label it a horror. To a moth, it would be a touching and dramatic love story.

After hatching from his silk cocoon, a male silk moth emerges only to find he is all alone, and soon discovers a horrifying truth—he has no functioning mouth! Since he cannot eat nor drink, he only has days to live. Nevertheless, he knows what needs to be done. His biology bids him to fulfill procreation, the primary goal of all living creatures. Luckily for him, male silk moths have been known to detect a female from over a kilometer away. How will he find his mate and make it there on time? We will find out with a set of experiments.

Experiment: Silkworm Moth Mating Behavior

In this experiment, we will investigate the sexual behaviors of adult silk moths. While live silk moths are not available to purchase in your average department store, you can readily buy living silkworm cocoons online. Once a cocoon forms, a moth usually emerges in 2–3 weeks. If you receive a batch of 10 together, they will all emerge within 1–2 days of each other. The first thing we will need to do is to identify the male and female moths and separate them from each other. No fooling around for these two!

When a moth emerges from the cocoon, quickly isolate it to its own jar so you can take a closer look at it. Count the number of segments in the moth's abdomen. If it has eight segments, it is a male. If it has seven segments, it is a female. Still can't tell? Females are generally larger, with bigger abdomens and wings than the skinnier male. The female also has a gland protruding from her back. As you identify their sexes, separate them into storage containers and keep them apart from each other. It's important that they not be allowed to mate!

Now it's time for the experiment. Find an open space to serve as your testing arena. Begin by placing two male moths in some plastic cups, and bring the cups close together, a few inches apart. Observe the moths and record your observations for 10 minutes. Next, in another set of cups, place some females, bring the cups together, and observe the pair of females. Finally, bring together a cup containing a male and another with a female. What do you notice?

In the male/male and female/female pairings, you probably saw very little movement in any of the cups. Both just sat there minding their own business. But when a male and female were in close proximity . . . watch out! The male starts to get terribly excited. His wings start vibrating, as he flutters and scatters about in seemingly random directions. This is what we call reproductive behavior, and we will use this behavior as a measurement for our next experiment.

Experiment: Silk Moth Chemotaxis

In the wild, silk moths can locate each other across long distances. But how do they track each other down? Sight or hearing are of no use when your love interest is so far away. In this experiment, we will determine what the nose truly knows, by investigating odors.

For this experiment, we will be using a chemical compound called bombykol, which can be ordered online. Take a small amount of pure bombykol (10 mg) and mix with 100 mL of mineral oil. You can store this mixture in a sealed bottle in the fridge for a while, if needed. Plain mineral oil will be the control solution. Both liquids smell about the same to us humans, but what about the silk moths? We can do a behavioral experiment to find out.

On a large surface, place a paper cup upside down on one side of the table and place a female moth on the top of the cup. The height of the paper cup should be high enough that the male silk moth cannot see the female on top. On the other side of the table, release a male silk moth. Watch for the direction of movement and observe if the male moth begins any courtship behavior. Do the experiment again, only now release a female moth.

Now replace the cup with a female moth on top with another upside down paper cup that has some mineral oil (our control) on the top. On the other side of the table, release a male silk moth. Watch for the direction of movement, and again observe if the male moth begins any courtship behavior. Also repeat this with the female moth.

Finally, we will replace the mineral oil cup with another upside down cup with a small amount of the bombykol and mineral oil mix from your sealed bottle. On the other side of the table, release a male silk moth. Watch for the direction of movement, and observe if the male moth begins any courtship behavior. Again, do the same experiment on a female moth. In fact, you should repeat this experiment with each of your male and female moths so that you can get a better sample of their behaviors.

So, what did you see? All the male moths should *zip* across the table to the base of the cup and start showing courtship behaviors for both a hidden female and for bombykol. But the females do not show any changes in behavior. Following is a graph of the male moths' responses.

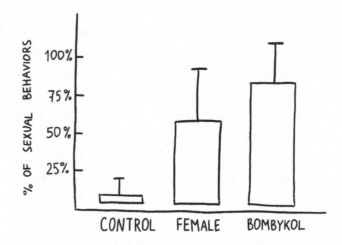

Notice that bombykol elicits an even stronger response from the male moths than an actual female moth! So how do we explain this quirky behavior? Bombykol is a special smell called a pheromone. Pheromones are a bit different from other smells. They are designed to send an exceedingly personal and specific signal, and when that signal hits, fireworks! Bombykol is a sex pheromone that a female silk moth excretes from her glands that attracts male silk moths. A very potent aphrodisiac, indeed! It allows for moths to find each other over long, long distances.

Experiment: Silk Moth Electroantennogram

In this experiment, we will determine how male moths smell bombykol. We have been able to determine that the moths can detect this chemical, but how does this pheromone signal find its way to the brain to help navigation? If you look at the silk moth, you will immediately notice incredibly large and feathery antennae. Could these antennae be sniffing for a lover? Let's do an experiment!

We want to electrically record neuron activity from the male moth's antenna. To do this, we anesthetize a male silk moth (in ice water) and cut off an end section of his antenna. Make an additional cut on the tip to expose nerves on each side. Using tweezers, gently lay the antenna across two recording electrodes, creating a bridge between them. Apply some conductive gel or paste to sandwich the ends of the antennae onto their respective electrodes, and we are ready to record! Note that you can make conductive gel by adding a tiny pinch of table salt to aloe vera gel.

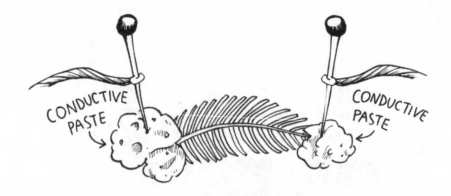

For the stimuli, we will soak some cotton balls in either mineral oil (control) or in bombykol/mineral oil solution. We will want to blow the chemicals onto the antenna from about three inches or so away. For consistency, you can use a small fan. Mark down when you exposed each cotton ball, and let's look at the electrical recording for each.

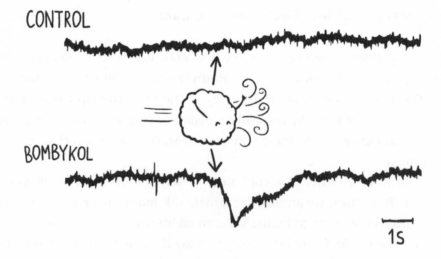

When bombykol was introduced, we saw a large electrical dip that wasn't there with the control. Something is definitely happening, but why doesn't this signal look like the spikes we saw in the cockroach leg? The answer lies in the way we are recording this signal. We are not recording from just a few spiking neurons. This type of recording is called an electroantennogram (EAG) signal, and it's the summation of hundreds of neurons in the antenna. This is why our signal looks so different. It isn't the discrete spiking of individual neurons we are observing, but the relatively slow change from a rest state to an active state of thousands of receptors and neurons in the antenna.

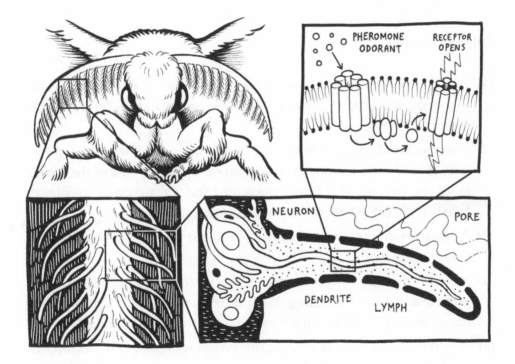

The moth's antennae are similar to our nose. They are used mostly for smelling and tracking scents from the air. The hairs on the antennae, called sensilla, are the key receiver for odor information. The bombykol interacts with specific neurons within the sensilla of male silk moths that will only respond to bombykol and nothing else. When an odorant binds onto a chemically selective protein, it opens a receptor, creating a larger voltage differential, which results in a spike. If enough spikes are firing, we can detect this using our EAG.

Your sense of smell works in a very similar manner. Scents are breathed in through your nostrils to a small patch of tissue high inside your nose. This area is filled with similar chemo-sensitive cells, called olfactory sensory neurons. Each olfactory neuron has one (and only one) odor receptor, and a human nose has around 400 types of these scent receptors. Each one is triggered by a different microscopic molecule. Once these neurons detect molecules, they send spikes to your brain, which then identifies the smell.

Follow-Up Questions

(1) What is the difference between a pheromone and an odor? Why would this make the behaviors we have seen so pronounced? Do you think humans have pheromones?

(2) Does the female react to bombykol in any stereotyped way? Perform an EAG of the female moth. How does the bombykol response of the female moth compare to that of the male moth? Did it match your hypothesis based on the behavior?

(3) Propose another behavioral experiment to test the courtship behavior in the silkworm moth.

(4) It is clear that the antenna responds directly to the bombykol, a process very advantageous for mating. What other "smells" could similarly help the silkworm moth? Do you see responses for these scents in the EAG?

(5) Why do you think understanding this system is important to neuroscientists? How could we apply these ideas to humans?

NEURONAL

ADAPTATION

7

Neuronal Adaptation

You may have noticed how hard it is to see in the daylight after emerging from a dark movie theatre after a matinee. It takes some time for your eyes to readjust to the bright sunlight. Also, when dressing for the day, you can feel your clothes on your body the moment you put them on, but rather quickly you forget the constant presence of cloth on your skin. If you stop and listen right now, you may hear noises (fans, appliances, ticking clocks, traffic) that you forgot were there. This is something the brain is really good at: it "learns" to adjust to various senses.

The mechanisms of learning are still a subject of healthy and active research. For example, scientists still do not know how declarative memories (like remembering the title of this book, for example) are stored in the brain. But we can chip away at these gaps in knowledge by conducting some experiments. Let's focus on this short-term effect of learning called "adaptation," the ability to tune out constant stimuli. One question you might ask is: where in the body does this process happen? Is it the brain itself, or can the sensory neurons do this on their own? Or maybe it's one of the neurons in the spinal cord? To start answering these questions, we will return to our old friend, the cockroach.

Experiment: Neural Adaptation

In this experiment, we will investigate how the brain adapts to unchanging stimuli using a cockroach. Our goal is to determine where stimuli learning occurs. The cockroach has a peripheral nervous system similar to ours. Sensory neurons in the limbs send their spiking activity to interneurons of the central nervous system living within the ventral nerve cord.

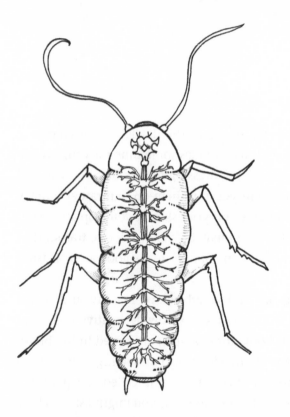

NOTE THAT
THE ROACH HAS:

6 ABDOMINAL GANGLIA

3 THORAXIC GANGLIA

1 SUBESOPHAGEAL
GANGLION IN ITS HEAD

1 "BRAIN" IN ITS HEAD

11 TOTAL GANGLIA

(HUMANS HAVE 31
IN OUR SPINAL CORD)

A cockroach only has 11 nodes in its ventral nerve cord (compared to the 31 expansions in our spinal cord), but the functional organization is similar. These central sensory neurons send information up to the brain. Each neuron in the chain could play a role in learning to adapt to constant stimuli.

We will start our investigation by looking at the primary sensory cells in the leg. We already looked at these neurons during our discussion of touch neurons in chapter 3, so we have an idea of how they encode touch information. Let's rerun the experiment, only this time we will apply long touches with constant pressure. In doing so, we will attempt to represent an unchanging stimulus that the brain should start to ignore.

So let's get started! Place the cockroach leg on a piece of cork, as we did in the original touch experiment. Position the electrodes close together so you can easily pick up spikes from individual neurons.

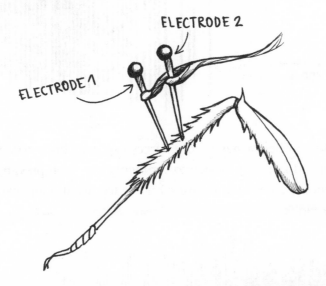

Turn on your SpikerBox, plug your cable into your smartphone or computer, and start recording. Find a barb that emits large spiking when touched.

Once you find a responsive sensory neuron, you will need to stimulate it with constant pressure for a long period of time. One way to do this is to place a toothpick on a manipulator or in a stack of books—something sturdy that you can move close to the leg. Next, position the toothpick over the sensitive spine on the leg of the cockroach, press against it, then leave it be. An alternative way to apply constant pressure is to cut a small piece of fishing line

(10 cm will work) and tape it to a popsicle stick. When you press it against the barb, the wire will bend and apply a constant force.

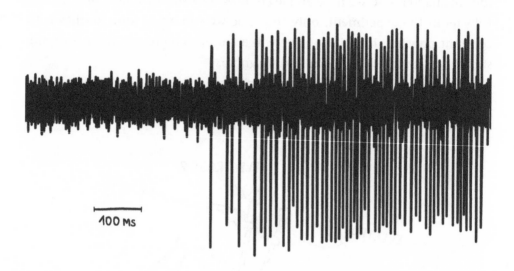

100 MS

You should now hear a massive discharge of spikes. This is no surprise, as we already saw this in previous touch experiments. But keep pressing . . . as the force continues at a constant pressure on the leg, what happens? Let's zoom out and take a look.

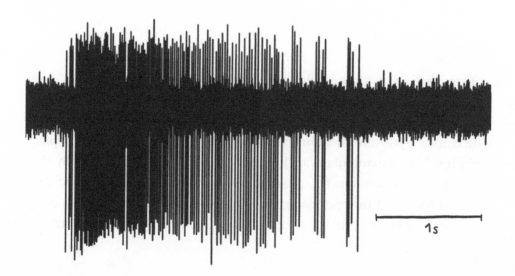

1s

The spikes start going away within a couple of seconds, and they diminish completely . . . even though a force is still applied to the touch receptor. The neuron is adapting to the unchanging force!

We can quantify this using the SpikeRecorder app. First, we must find the spikes to identify the time at which each action potential fired. Next, we can divide the period where we touched the barb into small windows of time (250 ms) and count the number of spikes that occurred during each one of these windows. This will show us how the spiking rate changes as we touch the barb.

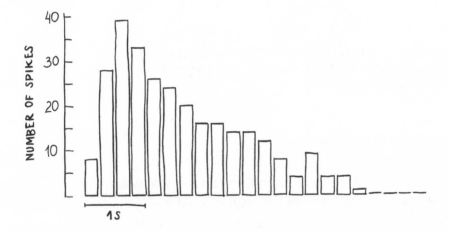

Here we can clearly see that the neuron is "learning" to ignore the constant stimuli. We can quantify the rate of adaptation by noting the amount of time it takes to go from peak firing to half of the peak. In this example, it took about 5 epochs, or 1.25s. Try this on a few neurons, and compare their adaptation rates.

This experiment shows us that even the very first sensory neuron can adapt. But does that mean that other neurons along the pathway (at the ganglia or brain level) do not play a role? No! To claim this, we will need careful experiments like this one to be conducted on the other sensory neurons. We can only say for sure that the first neuron is involved in learning, which is in itself surprising. It's not just our cortical cells of the brain that learn. This is happening all across our bodies!

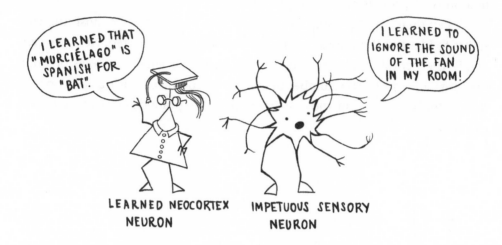

LEARNED NEOCORTEX
NEURON

IMPETUOUS SENSORY
NEURON

Follow-Up Questions

(1) What happens to the neuron firing after it adapts and once you increase the pressure? What is your hypothesis, and does your data match? Repeat the experiment, but lessen the pressure after adaptation.

(2) It's not only the barbs that are sensitive to touch. The exoskeleton of the leg itself (not the barbs) is sensitive. Try probing different parts of the leg, then measure whether the adaptation times are different.

(3) Why can't we do this experiment by hand? Try very precisely positioning the toothpick with your hand and measuring the adaptation time. It should appear to take much longer, if at all. Do you have any ideas about why this might be so?

(4) Do you think certain drugs will lengthen the adaptation time? Why or why not?

NEURONS FOR VISION

8

Neurons for Vision

Humans are visual creatures. We process complex visual information better than any other sensory signal. We can see the sunset with spectacular colors and observe the night sky bursting with stars. It's not just our eyes that perform this feat; our powerful brain does much of the processing, understanding, and admiring. It's our brain we have to thank for our remarkable ability to remember pictures. We can even recognize images displayed for only 13 milliseconds!

Look around your room right now. What do you see? Some walls, a window, a clock, perhaps. A plant, a table. It all looks so familiar and so real that it's easy to forget just how complex our visual system is. Light is bouncing off the world. Some light gets partially absorbed in some objects while being reflected by others. There are countless light vectors shooting off even dimly lit surfaces. Some of those light vectors pierce our eyes, setting off our visual process. In this chapter, we will decipher this code of vision. To do so, we are going to investigate an invertebrate with some real expertise on vision: bees!

Bees are skilled foragers that routinely identify the location of flowers, remember where they are, and communicate the food's whereabouts to others in the hive. They can travel great distances and precisely track down the flower with the most delicious pollen. But how does the bee see, and what does it see? Does it distinguish the same colors we do? To better understand how vision works, we will study the visual nervous system of this amazing creature.

Experiment: Electroretinogram (ERG)

In our first visual experiment, our goal is to detect whether there are electrical signals in the honey bee's compound eye. If the eye has neurons that respond to light, then we may be able to detect spikes when we expose the eye to flashes of light.

We first need to find a honey bee. You can seek one out near some flowers, bushes, or trees in your backyard. Wasps and bumble bees will also work. Capture one in a jar. To anesthetize the bee, you can put the jar in a refrigerator for a short while, or place it in a tub of crushed ice. Either way, when your bee becomes unresponsive, you can set it up for recording.

Find a room that can be darkened easily by closing curtains or doors. Take a piece of paper and place it on your work surface. Press a rolled piece of tape sticky side out onto the paper, remove your anesthetized bee from the jar, and gently stick the bee belly-side up onto the tape. Turn the tape over and press it onto the paper. The bee is now immobilized and ready for recording. You can use the paper to shuttle the bee back and forth to the fridge should you need to anesthetize it again between experiments. Fret not about its well-being! When the experiments are done, this method will allow you to remove the bee from the tape without hurting its wings. Once released, it should happily buzz away to its daily business!

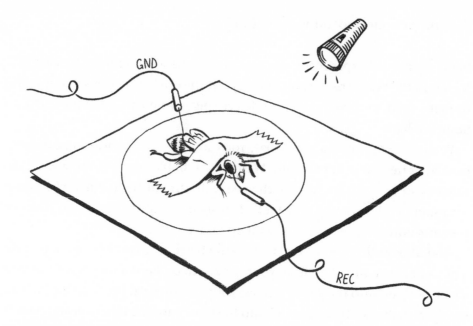

To prepare for recording, we will first set up the ground electrode. Using a small pin, slightly puncture the back of the bee's thorax (the midsection) and place a thin wire through the small hole. The location of the ground electrode is not that important, but it needs to stay within the bee during the experiment. Next, we will use another thin wire to act as the recording electrode and place it right on the facet eye as follows: bend or hook the recording electrode wire and press it up against the bee's eye, maximizing the contact area of the electrode onto the eye. Be careful not to puncture the eye! Apply some electrode gel on the electrodes to make sure you have good contact so we can improve our signal.

Everything is in place and we are ready to begin. Turn on your SpikerBox and start recording data. Cover the bee with a cardboard box and wait for a few minutes to let its eyes adjust. The box needs to have a hole on the side that the bee is facing. When you are ready, turn on the flashlight on your smartphone and blink a couple flashes of light at the bee, moving the light on and off of the eye.

Notice what happens to the electrical signal when the light is on the eye.

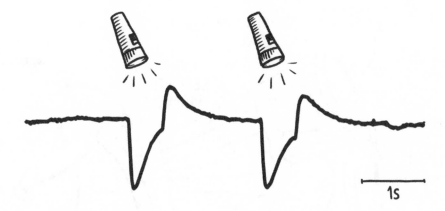

Something is happening! There is a dip in the electrical signal coming from within the eye. The light is being captured by light-sensitive cells within the eye and triggering chemical and electrical reactions, which are then processed by the brain. The electrode is measuring the conversion of photons into electricity by the bee's photoreceptors. We call this signal the electroretinogram, or ERG. While it might seem simple, this signal is the "hello world" of visual neuroscience, and variations of this simple lab can tell us a lot about how the visual system works.

This electrical activity of the ERG may look similar to the action potentials in the cockroach leg, but there are a few differences in the speed and shape of this signal. Here we are recording on the outside of the eye and are measuring the signals of photoreceptors, a special kind of neuron. They do not generate spikes, but they convert the number of absorbed photons into changes of electricity in a tonic fashion. Additionally, the ERG is the summation of electrical activity all across hundreds of hexagonal facets arranged in the honey bee's compound eye. These small facets, called ommatidia, tile the visual scenery of the bee into little pixels—like on your TV screen, but with a lower resolution. Photoreceptors within the ommatidia turn light into an electrical potential. The more photons they absorb, the larger the electrical change. The dip we see in the ERG upon visual stimulation is the collective response of the photoreceptors. Note the similarity in the shape of the ERG signal and the collective chemo responses seen in moth antennogram back in chapter 6.

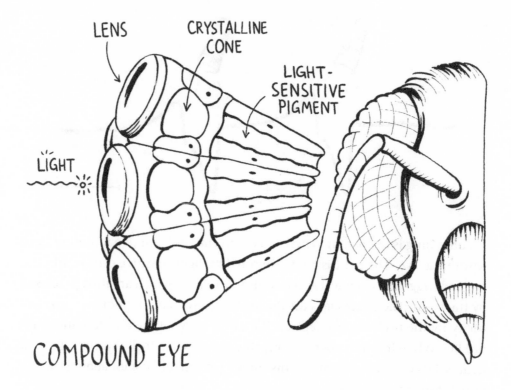

LENS   CRYSTALLINE
CONE

LIGHT-
SENSITIVE
PIGMENT

LIGHT

COMPOUND EYE

The overall waveform components of the bee ERG are fairly consistent across individuals of the same species. The signal might vary slightly based on your electrode type or location, and type of stimulus, but overall it should be highly reproducible. However, the ERGs of different organisms can often have dramatic variations, as visual systems are fine-tuned to their environments. One way that visual systems become specialized is their ability to detect specific and relevant colors. Let's find out what colors the bees can really see.

Experiment: Color Electroretinograms

Light consists of photons, elementary units that travel at the speed of light and behave like particles and waves at the same time. The waves of light are traveling in different wavelengths. It's precisely these varying wavelengths that allow us to perceive different colors of the spectrum. But what does the world look like from a bee's eye? In this experiment, we will test the bee's eye to determine whether it can detect different colors. First, we will investigate if it responds to the same colors as human vision using red, green, and blue.

We will set up the same ERG experiment as before, only this time we will shine colored lights on the bee. For this, you will need to collect 3 LEDs (red, blue, and green), or find some color film to place over your flashlight. When you are ready, turn off the lights and randomize which colors you shine onto the bee's eye.

Flash the colored light quickly on the eye to generate an ERG. Repeat in a random pattern with the other colors. You will start to notice an interesting pattern. Not all colors have the same responses. In fact, the responses are drastically different between colors, but very consistent between flashes of the same color.

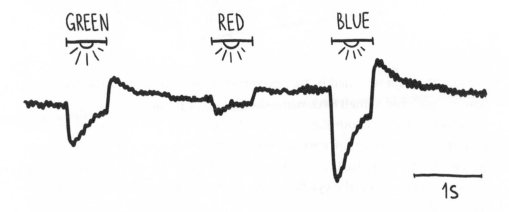

What is happening here? The photoreceptors in the bee's eye have evolved to absorb photons from only a small band of wavelengths. Our eyes possess, for example, three different types of photoreceptors that are sensitive to wavelengths of red, green, and blue. These photoreceptor types are being activated only by photons with wavelengths that lie in the red, green, and blue spectrum, respectively. The more suitable photons hit the photoreceptors, the more electricity is being generated, which in turn encodes the visual information of our surroundings into a language our brain understands.

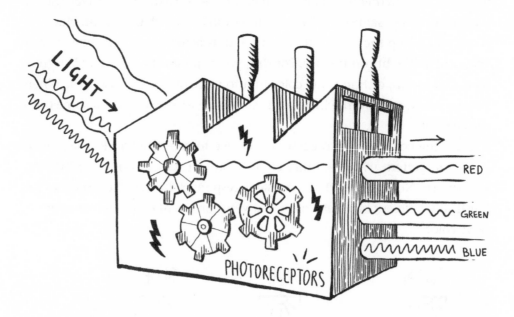

From the data recorded in this experiment, we can see that the red wavelengths produced a small ERG, while the blue color produced the largest signal. What does this say about the colors the bee can see? Red light must be close to invisible to the bee since we see little conversion from red wavelengths to the electrical activity that the brain needs. The blue is much stronger than the green, indicating that the eye has many photoreceptors that respond to blue wavelengths.

Experiment: Broad Spectrum Electroretinograms

We are now starting to get a picture of what colors a bee can detect. In this experiment, we will investigate this a bit more carefully by expanding the wavelength spectrum to include some colors of light that are invisible to the human eye—infrared and ultraviolet. We will set up the experiment just as before, placing an electrode on the eye, putting the cardboard box on top of the setup, and flashing colored LEDs. While small LEDs are easier to control, they require some electronics skill to set up. Alternatively, you can use your color-filtered flashlight, but you would need to procure an infrared and an ultraviolet flashlight (available online). Be sure to set up your device to record the electrode data, then randomly cycle through the five colors, stimulating the full eye for each one.

When you are done, we can sort through and find the average of the color flashes for each color. This allows us to see a clear picture of the response of the eye to the various colors. You can then determine the eye's color sensitivity by measuring the relative size of the peak deflection for each color. Arrange the average responses to the various wavelengths in order and measure the peak of each dip in the ERG.

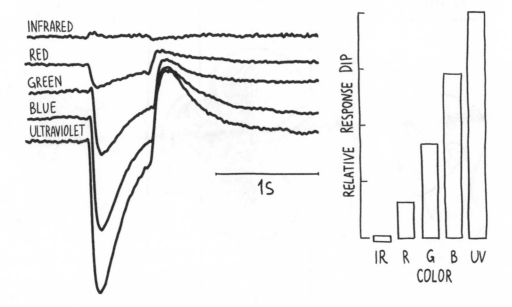

Incredible! From the data, it looks like the bees can see ultraviolet really, really well. This light is invisible to us, meaning they can see things we cannot! The bees can also see blue, green, and a bit of red. The infrared light, however, did not elicit a signal in our recording. The strong response to the blue light could explain why bees seem so profoundly attracted to blue flowers. But there is still one mystery: Why UV? What does it look like? And why are bees so sensitive to it?

It is not possible to imagine UV, as we are blind to the photons that travel in that wavelength. Humans have three types of photoreceptors, R-G-B; the bee has G-B-UV. The range of colors is more or less the same, but they are shifted. So, what do bees see? We can use cameras to help give us some clues. By converting the UV patterns on flowers into colors that are visible to us, we begin to notice a secret world. It seems that flowers make little airstrips for the bees using UV, guiding them to the sweet, yummy nectar and pollen.

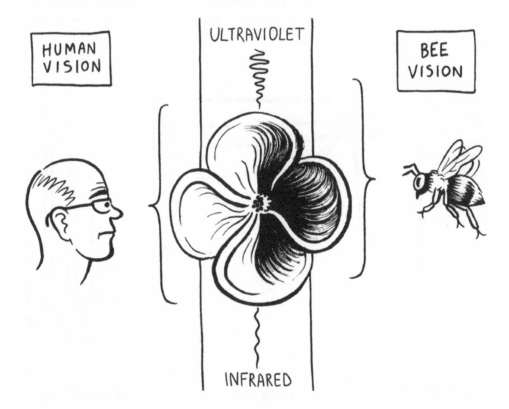

On the left, we have our boring human view of a flower, and on the right we have the bee's souped-up UV view. This is a beautiful example of how the bee's vision co-evolved with the color and patterns of plants designed to attract pollinators using visual cues.

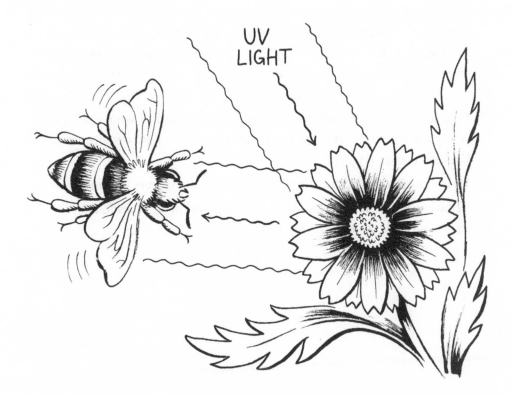

Color vision in bees was first discovered by Karl von Frisch (1914, Nobel Prize). He managed to show that bees can differentiate between colors and different shades of gray. Bees successfully remembered the colored pieces of paper after von Frisch put pollen on them and visited them more frequently, even in the absence of food. The ERG is an elegant way to confirm these experiments physiologically and to demonstrate the bee's trichromatic vision, including its difference from humans' visual perception of the world.

Follow-Up Questions

(1) TVs and movies send frames of still pictures at a speed fast enough to trick our brain into seeing motion. This frame speed that allows this illusion is referred to as the "flicker fusion" rate. Using the ERG, can you determine the flicker fusion rate of the bee? At what rate do multiple flashes of light look like one continuous light?

(2) Find another insect with an eye. Make a prediction of what colors it can see given where it lives, then perform your color ERG experiments. Did the results match your hypothesis?

(3) How would you perceive the world differently with eyes like an insect?

(4) What would the ERGs look like if you didn't allow the eyes to adapt to the dark? Why does it look this way?

NEURO PHARMACOLOGY

9

Neuropharmacology

We have carefully examined the various ways in which the nervous systems of different organisms interpret information using electricity. Spikes or action potentials have been the primary currency of information exchange, but there is still a lot more to discover about how neurons communicate with one another apart from electricity. During the late 1800s, scientists were grappling with the idea of neurons and were attempting to better understand the networks they created. The two leading theories contested whether or not the neurons were discrete (the neural doctrine) or fused together (the reticular theory).

Further investigations with extremely powerful microscopes would reveal that neurons are indeed distinct cells and not fused together. And thank goodness for that! If the brain were one giant neuron, our nervous system would effectively lack any sort of decision-making ability. If a signal was sent, it would continue to zoom throughout the nervous system indiscriminately. Instead, the nervous system is divided into tens of billions of small processing units called neurons, which communicate using a combination of electrical and chemical signaling. The crux of this communication is the synapse.

The synapse is a very small gap between axons and dendrites. This gap is around 20 nm wide (that's four thousandths the width of a human hair), which is so impossibly close that it was hidden from view until the scanning electron microscope was used in the 1950s. The synapse evolved to be small for speed. Spikes are fast and travel quickly down an axon, where it releases chemicals called neurotransmitters. But chemicals diffuse slowly, so the short distances can greatly speed up this communication process.

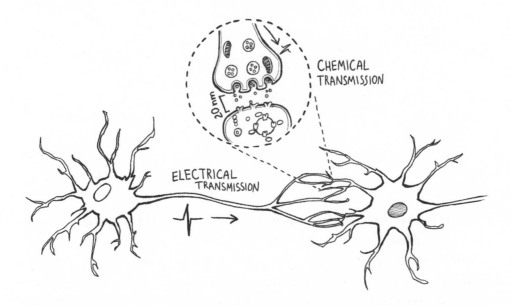

But what do the neurotransmitters actually do to the dendrites of the receiving cells? Do they increase—or maybe decrease—the voltage? In these sets of experiments, we will try to understand the roles of neurotransmitters.

Experiment: Cricket Cercal System

To investigate neurotransmitters throughout the central nervous system, we are going to be using a new model organism: the common cricket! Finding crickets is easy. You can catch them in tall grass anywhere you hear their characteristic chirp. Or, if you don't feel like chasing a cricket today, they are readily available at local pet stores as feeder insects for reptiles.

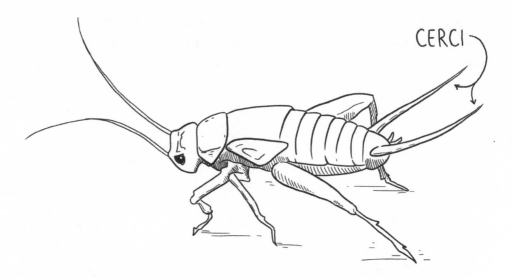

CERCI

Crickets have large sensing organs called "cerci" on the rear segment that are very sensitive to wind vibration. The cerci look like a pair of antennae growing from the posterior end. We will be recording from this cercal system in order to see the effects of various neurotransmitters.

Let's get started. The first step is familiar: place your cricket in ice water to anesthetize it. Use tape to secure your cricket down to a cork or balsa wood surface, but make sure not to cover the cerci. Pin your bioamplifier's electrodes along the central axis of the insect, with the ground near the center and the recording electrode near the cerci.

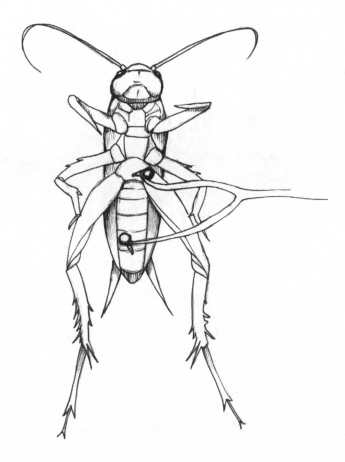

About 2–4 minutes will be enough for the cricket to "warm up." Then, blow gently on the rear of the insect. You should see the cerci move from the pressure of your air puff. You should also hear an increase in the spiking activity on your SpikerBox. These spikes are smaller, so they may not be as loud as the spikes you are used to hearing with the cockroach-leg prep. But if you prick up your ears, you should hear them well enough. Record a few sessions of blowing on the cercal system. This will be our control experiment.

We now know what the cercal response should sound and look like.

Experiment: Inhibitory Neurotransmitters

In this experiment, we will begin to test the effect of compounds on central nervous system neurons. We will bring a tiny amount of a chemical solution into the scene, and then look for changes in the activity of the nervous system when we stimulate its cercal system.

There are many drugs that affect neurons, but many can be illegal or dangerous to obtain (such as narcotics). We will save that research for well-funded professionals who can navigate all the associated bureaucratic red tape. Fortunately, as DIY neuroscientists, we can simply go to the grocery store and find many low-cost neuro-active chemicals! In this experiment, let's look at an ingredient used in cooking: monosodium glutamate, better known as MSG. MSG is best known for its incredible umami or savory taste. Once dissolved in water, it turns into positively charged sodium ions and negatively charged glutamate ions.

$$O \quad O$$

-O   OH

GLUTAMATE   NH₂

To create your glutamate solution, you need to find some powdered/crystallized MSG from the spice aisle in your grocery store. Fill up a clear pill bottle about a quarter full of MSG salt crystals, fill the remainder of the bottle with water, and shake thoroughly to dissolve the MSG. Note that not all the MSG will dissolve, so the product will be a saturated solution. Fill a syringe with a small amount (0.1 cc) of the MSG solution. We use insulin syringes for this experiment, which are available over the counter at any pharmacy.

Repeat our cricket cercal preparation from the previous experiment and record a few blowing responses. Next, take the syringe and inject a tiny bit of the MSG solution into the cricket near the cerci. Wait a few seconds after the injection, then listen and watch your signal. See if you can observe any difference in the firing rate without stimulation. Compare this signal to the previous control recording when the cricket was at rest and was stimulated before the MSG injection. Then blow on the cerci again.

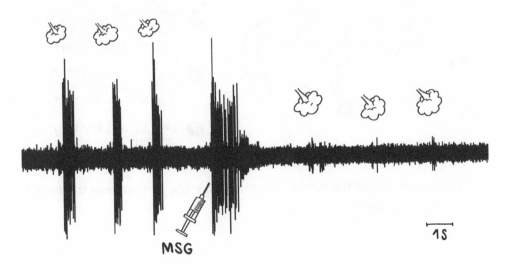

MSG

1 S

You should notice a difference, but it may not have been what you expected! In our experiments, the glutamate solution nearly shut down all the neurons from the cercal response system from firing. How does it do this? By binding to neurons via glutamate receptors in the synapse!

These receptors are one case in which the cricket diverges from the human system. Humans and crickets both use glutamate, but evidently it means different things to them. In humans, glutamate is a powerful excitatory neurotransmitter. Flooding a human synapse with glutamate will trigger sodium channels to open, creating an electrical difference more likely to fire an action potential.

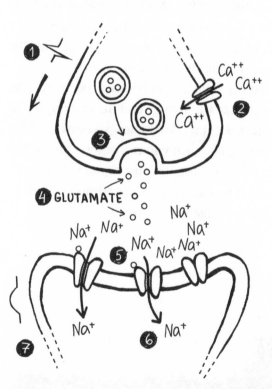

1 ACTION POTENTIAL REACHES END OF AXON

2 CAUSING CALCIUM TO ENTER DUE TO VOLTAGE SENSITIVE Ca^{++} CHANNELS

3 CAUSING SYNAPTIC VESICLES TO FUSE WITH MEMBRANE

4 CAUSING RELEASE OF GLUTAMATE INTO THE SYNAPSE

5 WHICH BINDS TO LIGAND GATED CHANNELS MAKING THEM OPEN

6 RESULTING IN SODIUM INFLUX AND A CHANGE IN VOLTAGE

7 CALLED A SYNAPTIC POTENTIAL

In fact, over 80% of the synapses in your brain use glutamate as their excitatory neurotransmitter.

However, we have just observed the opposite in our cricket. Adding glutamate to its synapses actually inhibits neural activity. In the vertebrate central nervous system, inhibitory channels are mainly controlled by a neurotransmitter called GABA. So, here we see a major difference between the human and insect. Both are using the same neurotransmitter glutamate, but it has completely different effects.

Experiment: Excitatory Neurotransmitters

In this experiment, we will attempt to inject a different pharmacological substance: nicotine, which is a natural stimulant that comes from the tobacco plant. We know that nicotine affects neurons, as the chemical is highly addictive in humans. So let's see how it affects the neurons of invertebrates. Would crickets start smoking if given the chance?

NICOTINE

To create your nicotine solution, take a cigarette or small cigar, remove the shredded tobacco leaves, and fill a small container with them (a clear pill bottle, for example). Then fill the rest of the container with water, put the cap on, shake up the mixture, and allow it to sit for a couple of days to extract the nicotine into the solution. Over time, the liquid should turn yellowish brown. Fill another syringe with a small amount (0.1 cc) of the nicotine solution.

Prepare the cercal system as described above and record with a few blowing responses. Then take the syringe and inject a small bit of the nicotine solution into the cricket near the cerci. Wait a few seconds, then watch and listen carefully. You should begin to hear a large increase in activity and see twitching in the legs of the cricket!

What is happening? Following the injection of the nicotine solution, we can visually observe a dramatic difference in the firing rate of the neurons in the cricket's central nervous system. The nicotine is acting like an excitatory neurotransmitter. The injected nicotine binds to nicotinic acetylcholine receptors in the synapse and makes them hypersensitive to acetylcholine. There is enough acetylcholine in the synapse to make the neurons fire spikes like crazy.

So why does the tobacco plant produce nicotine? This experiment gives a pretty good hint. The nicotine alkaloid is naturally produced in tobacco as a defensive chemical to repel insects. It is a neurotoxin for insects, but in humans, our nicotinic acetylcholine receptors are found mostly in the human nerve-muscle interface. So nicotine acts as a stimulant in our neuromuscular system!

Follow-Up Questions

(1) Why did we switch to the cricket cercal system? Could we have done these experiments in our cockroach-leg setup?

(2) What other chemicals that affect neurons could you use for testing the cricket's cercal system reaction? Think of drugs that affect the nervous system.

(3) Could you repeat the excitatory neurotransmitter with "vape" juices? These claim to have very specific concentrations of nicotine. You will want to consider the effects that the propylene glycol solution may have.

PART II

BRAINS

II
Brains

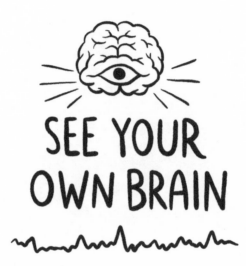

SEE YOUR
OWN BRAIN

10

See Your Own Brain

If you were to ask your friends and colleagues what exactly neuroscientists do, they would probably answer: "They study the brain!" We've already seen that neuroscience is much broader than the brain alone, encompassing entire fields dedicated to the sensory systems discussed so far in our model organisms. But there is no denying that many feel the brain itself is an ocean of mystery. And in this chapter, we begin to wade in and understand a bit more about how the brain works.

To truly grasp how something functions, we need to measure it. The brain is no exception. In our last section, we measured the electrical neural activity of model organisms by placing small wires close to axons of living neurons. Using instruments like scissors, pins, and sharp recording needles, we were able to spy on the communication of neurons as they processed information. Recording neurons in invertebrates is possible and even relatively easy for a number of reasons. First, the neurons in insects lack the electrical insulation layer (called myelin) that our neurons have. This lack of myelin makes the current we measure much larger and easier to record. Second, our model invertebrates by definition have no bones. They have an outer shell that is readily penetrated by sharp electrodes. Once inside, the electrodes can cozy up right next to the electrically active axons, neurons, and muscles.

Humans, on the other hand, have thick skulls protecting their brains. The most obvious way to record from neurons in a human is to drill your way through the skull and place the electrodes inside the brain. This is what often happens during brain surgery (using power tools to drill the hole). Place some wires in—and voilà, it works! You could hear spikes just like in our previous experiments. But you cannot do this at home, obviously, as the results would most likely be deadly (not to mention messy)! So you will need to find a way to record electrical signals from your own brain without having to poke, cut, saw, or drill your way past your skull. What about from the outside of the body? That seems to work in hospitals where the heart signals are monitored without direct access to the heart. What if we were to just place the electrodes on the outside of the skull above the brain?

Experiment: Electroencephalography (EEG)

For this experiment, we will need to connect our head to the SpikerBox. A simple way to do this is to place two metal snap rivets into a sports sweatband, and place the sweatband on your head like you normally would. The reason for a headband (besides being a good look) is that it will be easy to position it in different areas around the head, plus it's better to clip the metal leads from the SpikerBox to the metal buttons on the headband than to clip them to your skin directly. (Ouch!) Once you have your headband, position it so that the metal clips are just above your forehead.

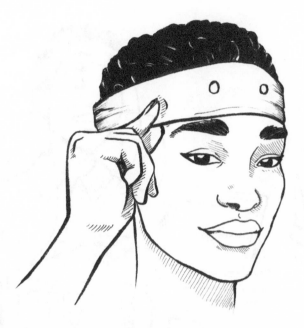

Next, place an EKG electrode sticker on the bone just behind your ear. This bone is called the mastoid process, and it is a nice, quiet spot without a lot of brain activity. Before we connect the headband to the SpikerBox, we will need to use a few globs of electrode gel underneath the metal clips so we can create an electrical connection between your scalp and the headband's electrodes. The electrode gel contains charged ions that allow any electricity on the head to flow more easily into the metal rivets. Once you are all set, you can clip the red recording clips onto your headband's metal clips, and the black ground clip onto the sticker electrode behind your ear. Connect all the leads to your SpikerBox.

It's time to start recording. Unlike in the cockroach labs from Part I, we will need to be a bit more careful with external noise in our system here. We recommend you perform your recordings on a battery-powered laptop, tablet, or phone—anything not plugged into the wall!

Connect your SpikerBox to your tablet or computer and open up the Spike-Recorder. Chances are you will see a flat, dormant line of inactivity, but don't let it frighten you! In all likelihood, you are not dead but rather have a bad electrical connection between the SpikerBox and your head. Try to add a bit more electrode gel and fiddle with the contact between your electrodes and scalp. If your connection is good, you will begin to see that the signal is ever so slowly moving:

100 ms

This may seem rather odd at first. It appears to be very flat compared to the other recordings we have made so far. This is not an error. Recordings of the brain through the scalp, skin, and hair are considered to be a very "weak" signal, compared to internal recordings. To get a better look, let's increase the gain in software.

100 ms

That's more like it! Now we can actually see that something is happening. Behold the electroencephalogram, or EEG! Notice that the EEG recording looks a lot more wavy than spiky. Maybe this slow-moving line is encoding our thoughts? Let's test this. Try thinking of two different things, back and forth. Do you notice any reliable differences between the two thoughts? It's best to envisage two completely unrelated notions, or different thoughts of stressful or relaxing situations.

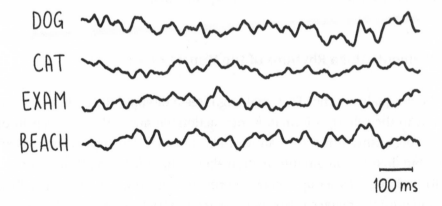

100 ms

It's hard to observe any recognizable patterns in these 1s snippets of the raw data EEG traces. Perhaps the EEG encodes the relative "strength" of the thoughts, like our neuron rate coding experiments of chapter 3. While watching your recording, think about the color blue. Think really hard! Think really, uh, soft?

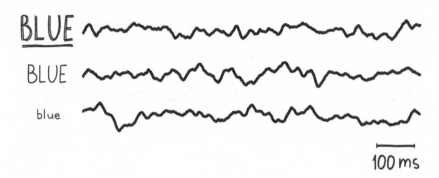

Hmmmm . . . There are a few things to note here. First, it can be difficult to perform human experiments on conscious thoughts. We will need to make better stimuli. Also, the EEG signals seem almost random—again! But how can this be? We are pretty sure that we are recording from the brain. Doesn't the brain control our thoughts? Shouldn't we be able to see at least some difference between thinking of objects versus activities? To understand what is happening here, we will need to do more experiments.

Experiment: Alpha Rhythms of the Visual Cortex

While the quality of the EEG signal might seem disappointing when you compare it to the robust, information-rich action we saw with the spikes in our model organisms, this is the nature of EEG research. The signals are weak and slow, but do they contain information about what the brain is thinking?

To find out, let's set up a new experiment. This time, twist the headband around until the metal electrodes are facing the back of the head. We are now aiming for our brain's occipital lobe. In EEG research, electrode positions are given identifiers to make it easier to communicate with other scientists about your experiments. These identifiers often refer to the region they are sitting over. For example, we will now be recording from O1 and O2 from the occipital lobe.

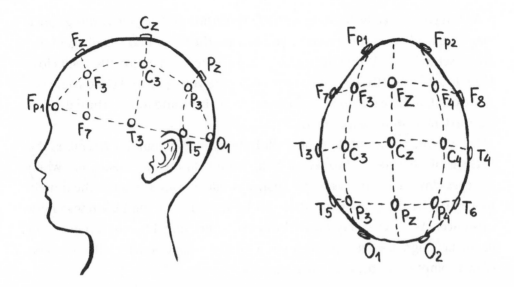

Just like before, we will add some gel to enable a good electrical connection underneath the electrodes. This time, make sure the gel seeps through your hair and onto your scalp. It should feel like a cool spot on your head when applied properly. If you have a lot of hair, you may need to have a friend part your hair and apply electrode gel to make good contact with your scalp. There is no need to shave your head for the sake of neuroscience—unless you really want to! Again, we will keep the ground electrode on the mastoid process.

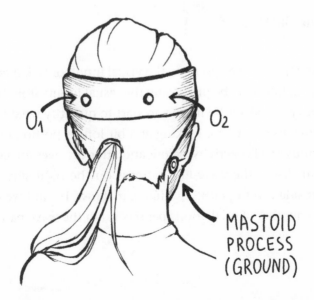

O_1 O_2

MASTOID
PROCESS
(GROUND)

When you are ready, hook up with the SpikerBox and begin recording. Once you have readjusted the screen so you can see the wavy EEG signal again, we are all set to begin. While sitting still, keep your eyes open for 10s, then close them for another 10s. Be sure to note each time you open and close your eyes so we can tell these instances apart when we go back and look at the data after we are done with the experiment.

So, what do the results look like? With the eyes open, our line resembles the EEG signal we were used to seeing from our previous experiments. But when the eyes close, a few things occur. First, you may see an artifact of the muscle movement in the eyelids (muscles can be the bane of the EEG researcher's existence). As the eyes stay closed, however, something different is happening in the EEG signal. There appears to be a lot more wavy activity with the eyes-closed compared to the eyes-open state.

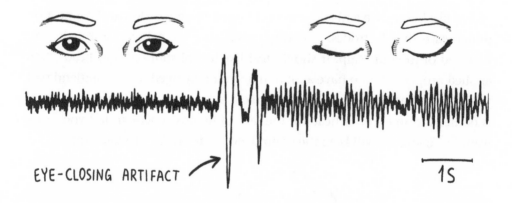

EYE-CLOSING ARTIFACT   1S

The EEG is changing in response to the visual input, or lack thereof! It seems that the occipital lobe may be processing the visual inputs from the eyes. Let's look at this wavy signal a little more and start to quantify what is happening. In the recording above, we are looking at a bit longer portion of time. There are a few seconds of EEG activity before and after the eyes are closed. Notice the increase in size of the wave-like patterns on the right side (eyes closed) versus the left side (eyes open)! The time bar shows 1s, and we can use it to determine the number of oscillations per second in the wave patterns.

Understanding frequency changes in EEG with time can sometimes be a challenge. So let's use a simpler wave to get used to the concept. Below are two sine waves:

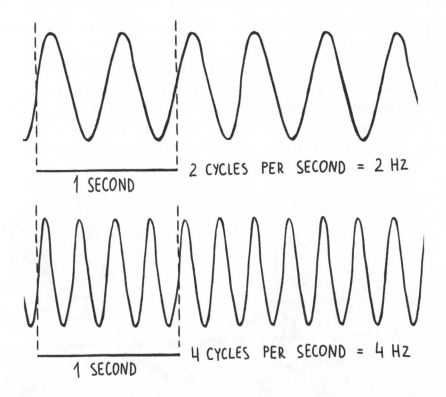

You can see from the 1s time bar in the top trace that it is oscillating at around 2 cycles per second (2 peaks or 2 valleys per second). The bottom trace is oscillating at 4 cycles per second. Let's imagine a signal change at 4 Hz and 2 Hz:

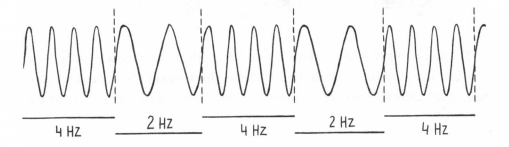

You can see that the signal is changing at 4 Hz and 2 Hz. Again, a quick way to note a dominant frequency in the EEG is just to count the number of peaks or valleys in a one-second window.

So let's go back to our eyes-closed recording and count the number of downward peaks in the EEG oscillations starting from the beginning of the time bar to the end (you can use your pen to touch and count out each peak). There are roughly 10 oscillations in that 1s time window, and we write 10 per second as "10 Hz" (Hz = 1/s). This 10 Hz frequency of the brain oscillation is called the "alpha" wave, and our data is showing that the power (height) of the alpha-wave oscillations increases when the eyes are closed. This 10 Hz rhythm was first discovered (along with EEG itself for that matter) in a very unlikely way.

The German psychiatrist Hans Berger (1873–1941) suffered a frightful military horseback-riding accident in the 1890s. This wouldn't have been much to note if his sister didn't telegram him from miles away expressing that she had recently felt a sudden feeling of fear for her brother's safety. This coincidence seriously baffled Berger, so he set out to explain how this "spontaneous telepathy" could work. How could he have used it to convey information about his health to his sister? To get to the bottom of it, he performed a series of careful electrical recordings from the heads of human subjects.

Berger developed a tool to record the electrical current of the brain from the surface of the scalp using rubber bands and silver foil—not very different from the sweatbands and buttons you're currently using! He was also the first person to describe the different waves or rhythms that were present in the brain. While he failed to determine a telepathic medium (the signals he found were far too weak to travel a great distance), he did make a historical breakthrough along the way. This episode made him the first scientist ever to discover that you could record electrical activity from the brain using the electroencephalogram. The first oscillating wave he discovered was the 10 Hz rhythm from our visual experiment, which he dubbed the "alpha" rhythm.

But why does this alpha rhythm happen when our eyes are closed and not open? The answer lies with how "synchronous" the brain activity is. The more synchronous the neurons in your brain are, the less data processing is occurring. This leads to the paradox that the stronger the electrical signal we can record on the surface of your scalp, the less interesting things your brain is doing.

When our eyes are open, the neurons in the visual cortex get to work by processing rich information streaming in from the retina. Given the complexity of vision, there is little activity that is synchronous with other parts of the visual field. When you close your eyes, your visual cortex stops receiving complex information from your eyes. The pyrotechnics of visual information are now replaced with darkness. The party's over, and now the visual cortex neurons are all doing the same thing at the same time—waiting for input. While at it, they are processing darkness, in an idling mode of synchrony that shows in our records as the alpha wave.

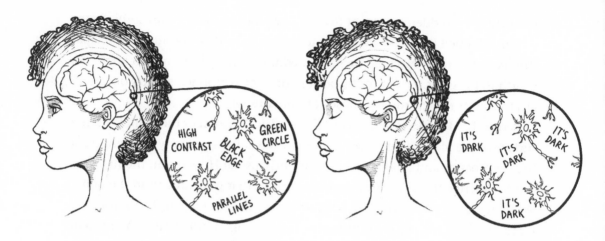

To understand why only synchronous activity can be seen in the EEG, let's do a thought experiment. Imagine a stadium during a baseball game, where everyone is having conversations between themselves. If you are outside the stadium, all you hear is a formless hum of noise. There are many interesting things being said (probably), but you can't detect or discern them from outside the stadium. This is similar to our EEG. Outside of the skull is too far away to listen in on the conversations of individual neurons.

Let us now tweak this concept to imagine all our sports fans in the stadium doing the same thing, like singing a national anthem. We can certainly hear the song, though distorted, outside the stadium, and know that the game is starting soon. This is analogous to the slow waves your brain generates while in deep sleep or the alpha waves the visual cortex generates when your eyes are closed. The crowd of neurons are all doing the same thing at the same time (idly waiting).

The loudest event we can detect outside the stadium is when the home team scores, as a huge population inside the stadium screams as one, very loudly, at exactly the same time. When large populations of neurons in your brain do this (all firing action potentials at the same time), it is called epilepsy, and it is very dangerous, but very easy to see within the EEG.

So to make an oscillating EEG signal, neurons need to fire synchronously in the same region, at the same time. Given the way our brain is structured into columns, there is an orderly arrangement of neurons, all aligned together in a similar direction. When these neurons receive incoming excitatory spikes, their internal voltage increases slightly. This electrical change is called the excitatory postsynaptic potential (EPSP). These voltage increases are very small (1–4mV) but can last 15–20ms before dissipating.

By itself, an EPSP is too small to be detected on the surface. But if there are enough of them occurring around the same time and in the same place within the brain, their voltage can amplify greatly. In fact, it can grow so large that it can be recorded on the outside of the skull!

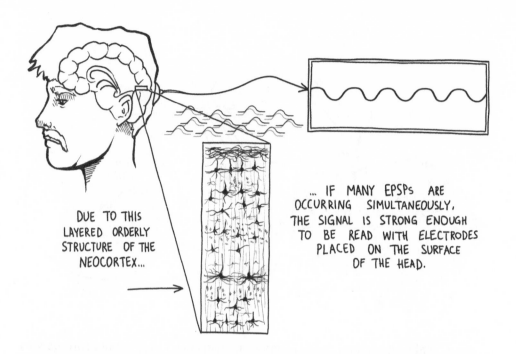

DUE TO THIS LAYERED ORDERLY STRUCTURE OF THE NEOCORTEX...

... IF MANY EPSPs ARE OCCURRING SIMULTANEOUSLY, THE SIGNAL IS STRONG ENOUGH TO BE READ WITH ELECTRODES PLACED ON THE SURFACE OF THE HEAD.

These accounts serve to show how even a full century can change surprisingly little. We are still interested in many of the questions that puzzled brain scientists during Hans Berger's time, and we still use some of the same methodology. But with these foundational skills, we can begin to perform many more simple EEG experiments that can help us better understand the nature of the brain.

Follow-Up Questions

(1) We say here that the alpha rhythms really don't reflect what someone is thinking, but is that really true? Try recording alpha rhythms from a person while they're thinking different thoughts. Does it have any effect? Why or why not?

(2) What effect do certain conditions and factors have on aspects of the alpha wave? For example, a person's age and sex, how much sleep they've had,

whether they've had caffeine recently. Why do you think these do or don't have an effect?

(3) Try recording from different parts of someone's occipital region. Is there an optimal placement for seeing alpha rhythms? How far from this placement do you still see them?

(4) Try recording from different parts of someone's scalp. Do you see any other kinds of waves? What is the frequency? Are there changes in activity in other parts of the brain during the eyes-open or eyes-closed conditions?

11

Sleep

Sleep is a crucial part of our everyday life. We spend one-third of our lives asleep, and it's a phenomenon that is so integral to our daily lives that we often take it for granted. And while you may not know exactly what happens when you sleep, you probably know a lot about what happens when you don't sleep. Pulling an all-nighter while traveling, working, or studying can dramatically affect your mood and alertness the following day.

Sleep is a part of the daily cycle that is orchestrated by our brains. To get a better idea of what our brains are doing during sleep, we can attempt to take a look at the brain waves that occur during a good night's sleep.

Experiment: Sleep-Cycle EEGs

For this experiment, you will need to ask a friend to volunteer to take a nap! But before they go to sleep, you will need to attach a headband above the part of the brain called the frontal cortex. This is located toward the front of our skull, behind our forehead. Connect a ground to a bony spot just behind the ear, and clip the leads across to the electrodes on the headband. Using the EEG standard location map introduced in the previous chapter, we will record from electrode locations Fp1 and Fp2 (Fp being shorthand for prefrontal). You should now be ready to record electrical signals from the headband again. If it doesn't look like you are getting a good connection, you can add some more electrode gel between the metal electrodes and the skin.

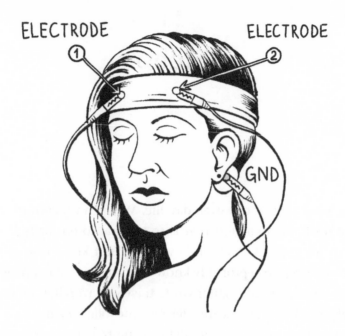

Now that you've got the cables connected, it's time to have your volunteer catch some z's. Have your friend lay down comfortably on a couch, and set your equipment on a coffee table nearby. The electrodes are positioned on the forehead, so if your friend can fall asleep while lying on their back, you should be all set. The EEG headband works best for naps, as overnight sleep involves body movements that shift the setup. Set an alarm for at least 90 minutes. Most people can't fall asleep immediately, so you should add another 10 or 20 minutes, just to be safe. While it is possible to do this experiment on your own, we do find it easier to record from another subject so you can observe the signals while they are sleeping.

Plug the electrode cable into the SpikerBox and begin recording. Observe your friend as they fall asleep and take notes. Do you notice some large deviations in the EEG while they are getting settled in? These are most likely movement artifacts. Just wait some time, and watch the results stream by your screen. As your subject lies still, the EEG signal will begin to flatten out.

AWAKE   1 S

As they try to sleep, you may start seeing some ripples appearing on your screen. When you see a particularly nice ripple going by, pause the display so you can count the number of peaks that occur in a second. This will give you a rough idea of the frequency you are seeing.

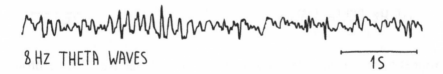

10 Hz ALPHA WAVES   1 S

Here we are seeing alpha waves, or about 10 cycles per second. When you see this, there's a good chance that your partner is still awake, but drowsy and getting close to falling asleep. Keep letting the data scroll by. It's subtle, but you may notice the waves getting a bit larger in amplitude and a bit more spread out in time.

8 Hz THETA WAVES   1 S

These oscillations are around 8 cycles per second and are referred to as "theta waves." These waves (and more alpha waves) will come and go for about 5–10 minutes. This is considered the first and lightest stage of sleep (Stage N1) and is also referred to as "sleep onset." It is characterized by these theta (4–8 Hz) EEG rhythms as well as slow eye movements lasting a few seconds. People who wake up from this stage may claim they weren't even sleeping . . . just resting their eyes!

As you keep watching the traces, you may see a few spots where the signal oscillates a bit faster.

15 Hz SLEEP SPINDLES   1 S

These are called "sleep spindles," and they oscillate around 12–15 Hz. This is a good indication that your subject has slipped into yet a deeper stage of sleep (Stage N2). This stage is characterized by very few eye movements, and the EEG signal is dominated by these sleep spindles. These sleep spindles will return about every 1–2 minutes during phase 2, until suddenly you begin to notice something strange in the signal. The waves are now much larger and oscillating much more slowly!

0.5 Hz DELTA WAVES   1 S

Your subject is now entering slow-wave sleep (Stage N3). This is probably the most identifiable signal of the sleep stages, thanks to these slow-moving delta waves. These voltages are typically much larger than the other rhythms, and are between 0.5 and 4 Hz. They are easy to spot, as they are the strongest synchronous activity the brain produces (outside of neurological disorders like epilepsy). Slow-wave sleep (SWS) is when your body really settles into slumber. It is the deepest stage of sleep, the one that is the least responsive to outside stimuli, and the hardest stage to be woken from. As you continue to record, you will notice that these delta waves disappear and are replaced by very minor fluctuations with no discernible patterns. It looks similar to the EEG from when your subject was still awake!

REM   ——————
                                                          1 s

But you can tell that the subject is still sleeping, with their eyes closed. You may also be able to see that their eyes are moving underneath their eyelids. This is the rapid eye movement (REM) stage of sleep. This stage is the most intriguing part of sleeping, as it is here that you have vivid dreams. It is also the stage of sleep with the most intense and diverse brain activity. Because the brain is so active, the REM EEG often looks like an "awake" EEG composed of a mix of various high-frequency, low-amplitude components.

You may be surprised to know that REM wasn't formally discovered until the mid-twentieth century. Even more curiously, this discovery didn't happen in a lab, but on a train! Scientific discoveries can happen anywhere.

In 1950, British physicist Robert Lawson noticed while on a long train ride that people's closed eyes were twitching while they were sleeping, and that this eye twitching would abruptly stop and restart at some point later. Being an observant and curious scientist, he continued to take note of his observations, and later published his account in a short letter in the scientific journal *Nature*. Talk about a productive train ride!

Experiment: Sleep Hypnograms

Keep recording your friend for as long as you can. See if you can recognize the various stages we've observed (N1, N2, N3, REM) as they begin to appear again. Write down the time in which the stages change. You should review the EEG stages from the recording (after you've taken a nap too, of course)!

The process of identifying the sleep stage given the EEG is called sleep scoring. This scoring method was first described in detail in a 1957 paper by researchers at the University of Chicago, graduate student Bill Dement and his advisor, Nathaniel Kleitman. They proposed the four stages of sleep described above, and their schema is still used in the present day with surprisingly few alterations over the years. Each sleep stage tends to follow a pattern cycle that lasts about 90 minutes.

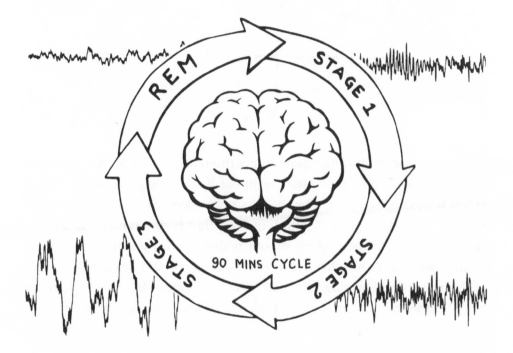

After you've scored the EEG recording, you can visualize the different stages as a function of time in something called a "hypnogram." Draw the horizontal lines indicating the start and length of stages you noted at the appropriate sleep stage, and draw vertical lines to connect stages together. In our idealized hypnogram below, you can see that more REM occurs as the night draws on and less time occurs in deep sleep.

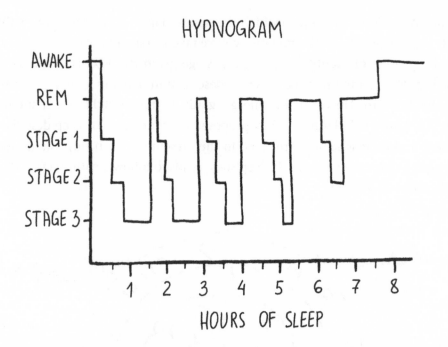

HYPNOGRAM

HOURS OF SLEEP

Artists and scientists alike have long written about this strange phenomenon that plays such a profound role in our lives and our health. Our fascination with sleep is matched by the mystery of sleep: Sleep itself is not very intuitive or easily explained. Think of our other bodily functions; they all have direct links to their effects on our bodies. We eat for nutrition and to get stronger, and we breathe to oxygenate our blood. And when we are tired, we sleep, and thus ensuring . . . what? Why do we need sleep? Well, there are many theories about why we sleep. Let's take a look at four key theories.

Protection Theory

Sleep keeps us safe. According to the first theory, we sleep to protect ourselves from things that would hurt us. Many animals are inactive at night for survival, burrowing away to safeguard themselves at their most vulnerable. Darkness provides natural camouflage that their predators can't see through. Wild chickens will fly into the trees at night to roost when it is dark. Other animals dig holes or build nests out of harm's way. A burrowed field mouse has a very real barrier between itself and its owl predator. But predators wouldn't be nearly as efficient in their gory job if they weren't adaptable. Many big cats slack away their days so they can fully activate at night—precisely when their supper is inactive.

Energy-Conservation Theory

Sleep conserves energy. This theory posits that the primary function of sleep is to conserve our energy resources, especially at a time when it does not make sense to utilize them—for example, at night when many animals cannot see to effectively hunt or gather food. Competition for food and energy resources is part of natural selection, and sleep provides a period in which the body's demand for energy is low. During sleep, energy metabolism, body temperature, and caloric demand all decrease, lessening the strain on the body. Animals like bears and squirrels are this theory's biggest advocates. They even take it to an extreme, going into a deep hibernation sleep that lasts the entire winter.

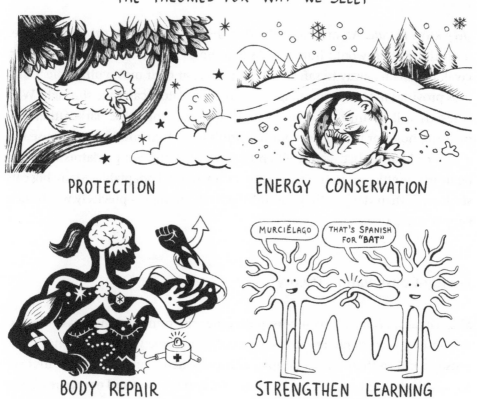

THE THEORIES FOR WHY WE SLEEP

PROTECTION

ENERGY CONSERVATION

BODY REPAIR

STRENGTHEN LEARNING

Body-Repair Theory

Sleep is a time for rejuvenation and repair. The third theory is that sleep allows our bodies to repair themselves. In effect, sleep lets us recharge our batteries for our next endeavor. Remember those countless times your mom told you something like "Rest up, tomorrow's a big day?" This is the theory she had in mind. Indeed, our body loses immune functions without sleep, leaving us more susceptible to stress. Without sleep, our metabolism doesn't work as well either, and we are at risk of injury due to extreme lack of focus. It's unclear whether sleep deprivation alone can kill you directly, but it is undisputed that it weakens your body and puts you at a much greater risk of sickness and injury. Sleeping regularly, however, enables muscle growth, tissue repair, protein synthesis, the release of growth hormones, and the facilitation of metabolic and hormonal processes.

Strengthen-Learning Theory

Sleep promotes learning. The fourth theory is that sleep helps us become smarter. While your body is resting, your brain is busy processing information from the day and forming new memories. For memories to be useful, three functions must occur: (1) acquisition—you must learn something new; (2) consolidation—the memory must get its own storage bin in the brain; and (3) recollection—you must be able to access the memory in the future. When we sleep, the theory goes, our brain is performing consolidation, strengthening the neural connections of experiences we had throughout the day and storing them as long-term memories. This theory seems plausible to anyone who has pulled an all-nighter cramming for a test. Without adequate sleep, your brain becomes foggy, and that figure you knew so well at 2:00 a.m. could not be recalled the next day.

How much sleep we need is largely dependent on our age. Infants typically need 13–14 hours of sleep, and the effects of sleep deprivation impact their learning capacity as they grow into adulthood. Adults, on the other hand, need less sleep, according to experts (with recommendations ranging from 7–9 hours per night), but sleep can still help strengthen memories and learned motor skills.

Sleep and its purposes have fascinated human beings since early human history, with documented sources dating back to Mesopotamian cultures. With this experiment introducing you to brain waves during sleep, you too can join the cycles of inquiry.

Follow-Up Questions

(1) Which hypothesis on why we sleep (protection, energy conservation, body repair, and strengthen learning) do you find the most compelling? Why? Can you design an experiment testing your reasons?

(2) You may have noticed that the EEG is much stronger and more detectable in the delta waves of sleep than during awake EEG. Why do you think that is?

(3) You can notice your dog's REM sleep. When, during sleep, your dog's eyes and mouth are twitching, with an occasional tiny quick leg movement, your dog is in REM sleep. This of course leaves any curious mind to wonder what it is that dogs even dream about. Maybe you can repeat Lawson's famous train-ride experiment and simply observe your dog when it begins sleeping and write down the time and duration when its eyes start twitching versus not twitching.

(4) Stage 2 (N2) sleep is often defined by sleep spindles preceded by a single large deflection known as a "K-complex." This K-complex is not very pronounced in the frontal location described here. Can you move the electrodes to another location to find it?

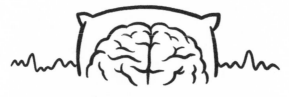

ENHANCE
YOUR MEMORIES
DURING SLEEP

12
Enhance Your Memories During Sleep

It can be so hard to remember things! Facts, names, where you left your keys, whether you locked your door or turned off the stove. Some people find it difficult to remember things easily. But did you know that memory, like muscles in our body, can be strengthened and enhanced? We learned in the previous chapter that one of the theories of sleep is that it helps reinforce memories and acquired skills. Perhaps we could take advantage of that knowledge and ditch the flashcards, instead trying an interesting way to hack our memory while we sleep! Let's put this theory to the test.

Experiment: Targeted Memory Reactivation

In this experiment, we need memories that can carefully be controlled and their recall accurately measured. To do this, the subject will use our app called TMR Memory Test from the App Store. This game will hide the location of pairs of objects on the screen, and the goal is to remember where they saw each pair. This is similar to the memory game you may have played as a kid, but with one catch. Each image in the memory game will have unique sound effects associated with the clue (like a car horn when you match the pictures and locations of two cars). Our hypothesis is that if we play just these sound effects for the subject later when they are asleep, the cues will help the sleeping subject selectively strengthen the cued memories. We will only play back half of the cues, and leave the uncued memories as a control. This experiment will be divided into 4 phases.

Learning Phase

The first phase is simple: have the subject play the memory game app to learn the positions of the cards. First, the subject will match picture cards on a grid all faceup as they begin to store the spatial and content memories in the short term. Next, they will do more rounds where the cards are facedown, and they will be asked to match pairs, each featuring a distinct sound effect that plays when the user matches the cards. After a few games of the same cards, the subject should have mastered the positions. You can experiment with how many times you will let the subject play through the learning phase of the memory game, but 2–3 times is enough for now.

Pre-Sleep Phase

Shortly before you begin the sleep phase of the experiment, have your subject play the memory game. The app will score their results and collect this data for later comparison. Ask a few qualitative questions: Were any of the matches easier to remember than others? Were they better able to remember the position or the content of the pictures? Then get them ready for the sleep phase.

Sleep Phase

Set the subject up for frontal cortex EEG sleep recording exactly as we have done in the previous chapter. You can let them sleep overnight or, even better, have them take a 90-minute nap. As the subject starts their nap/sleep, you need to monitor their EEG recording and detect when they transition into deep sleep/slow-wave sleep. You will know you are in the right phase once you start spotting delta waves in the EEG signal. Like we saw in Stage 3 sleep previously, delta waves are large and slow oscillations in the EEG lasting 1–2s per cycle. Keep waiting until they show up.

NO DELTA WAVES

DELTA WAVES!

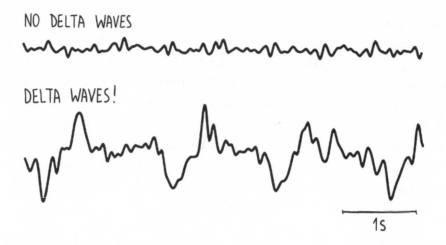

1s

When you detect delta waves, the subject is in slow-wave sleep. Wait some time for the delta waves to stabilize, then use the app to begin cueing the sounds from the game. The app will only be providing cues for half of the memory tasks. That way, after the subject wakes, you can see if the cued versus uncued pictures/locations had better recall.

START SOUNDS HERE

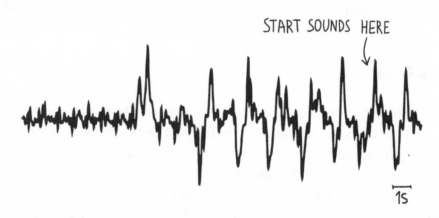

1s

When the subject wakes up, end the recording and save the EEG session for later reference. Ask the person if they heard, saw, or felt anything while asleep. If they answer yes, then they probably weren't in deep sleep. You can still run the rest of the experiment, but make a note of this.

Post-Sleep Phase

Have the subject play the memory game again and hand back the device. All data for the pre-sleep and post-sleep game sessions were captured by the app, allowing you to perform quantitative analysis. Specifically, we want to measure how many pixels away they were from selecting the center of the correct target location. Ask the same qualitative questions. Were there any differences in their answers from pre- and post-sleep?

Let's look through the quantitative results. We want to compare the average distance from the answer to the correct answer. We will take the post-sleep distances and subtract them from the game the subject played just before they went to sleep. How did the cued versus uncued memories compare?

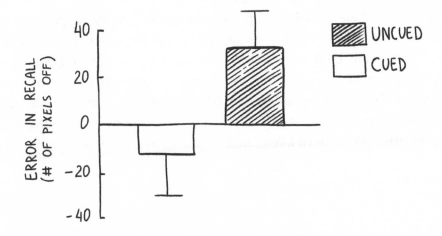

In this calculation, cues that were forgotten during sleep become a positive number (errors were greater after sleep than before), while cues that were better recalled will be a negative number. In our trials across five different people, we found that the cued memories were better remembered than the uncued memories! This seems almost impossible, but it matches the findings from other contemporary scientists.

What Makes a Memory?

There are many theories about how memories are stored and recalled, but most researchers agree that episodic memories depend on multiple regions of the brain's cortex. Episodic memories are the memories of times, places, and emotions, including the who, what, when, where, and why. For example, you may remember a wedding you went to last year. You can recall the ceremony, how the reception hall was decorated, what the band sounded like, characteristics of the speakers' voices, which friends and family were there, the emotions you felt. These many features of the event would be represented in multiple regions around the cortex that are specialized for processing different sorts of information. Declarative memories are built on representations that link across these various elements.

Storage of these feature-rich memories depends on the hippocampus, a seahorse-shaped brain structure located just under the cortex. The hippocampus, whose name derives from the Greek word *hippokampus* (*hippos*, meaning "horse," and *kampos*, meaning "sea monster"), plays an important role in the consolidation of information from fragile short-term memories into stable long-term memories. It does this through its connections between many different cortical areas, much like a hotel clerk assigning guests to their rooms.

This cross-cortical storage requires the connections of the hippocampus after initial learning. After memories have undergone a sufficient amount of cross-cortical consolidation, the memories can become stable. This means the neurons of the hippocampus are no longer required to invoke the set of distinct cortical networks attached to the memory. These memories can be retrieved again if enough associated cues are activated for a recall.

So why does cueing during slow-wave sleep actuate memories? This may have to do with something called "reactivation." Memories are not static, nor are they perfect. Rather, they shift and change over time as they are recalled. Reactivation is when something (for example, a place, voice, or smell) triggers a recollection of a memory. Sometimes this recall can strengthen the "truth" of the memory, while other times it can do the opposite. For this reason, there is a growing body of evidence against using memories as evidence in court cases. We also know that our episodic memories are accessed during sleep. Dreams often contain fragments of recent events stitched together with bits of other things we know, allowing for these memories to be stored for the long term. As studies of rats show, neurons that represent recent memories are literally "replaying" those memories in the hippocampus during SWS. What if we could activate a part of a memory during SWS? Since the audio cue is associated with the tile-location memory, we could reactivate the entire memory and strengthen it. In addition to reactivating specific associations using sounds, some studies have shown that learning sessions could be enhanced using odors during slow-wave sleep.

Now that we've observed one way we can "hack" our sleep to improve memories, we've unlocked the door for further studies. The tricky part seems to be a creative issue. How do we cue different kinds of memories during sleep? We're excited to see what you will come up with!

Follow-Up Questions

(1) We cued memories in Stage 3 sleep (slow-wave sleep), but what if we played the cues during other stages of sleep? Try the experiment with REM or another sleep cycle.

(2) Imagine and describe a consumer product designed to help you learn faster while you sleep. What are the challenges in making this a reality?

(3) What would happen if we changed the cueing percentage to 0%, 25%, 75%, or 100%?

(4) Does running the experiment during the night yield better results? Will having subjects perform this study before sleeping normally at night produce different results than having them nap in the middle of the day?

(5) How do surrounding environmental factors (light, temperature, space, outside noise, etc.) affect the performance of the subject?

DEALING WITH THE UNEXPECTED

13

Dealing with the Unexpected

Our brain is constantly taking in and processing information from the world around us. Many of the sights and sounds traveling to the brain are familiar and not very useful, so they can be ignored. This neural adaptation is very helpful as it frees us to focus on information that is important or new.

Many of the decisions we make require us to filter out an ocean of alternative options, in order to find a particular choice. Think about shopping online: you can scroll mindlessly through dozens of pictures until you find the item you are looking for. Or imagine sitting at the DMV for your assigned number to be called. You don't quite know when your number will show up (or when your product will appear), but when it does, your brain is ready. When you finally hear your number called, you may experience a small "Aha!" feeling come over you ("That's me!"), and you then quickly collect your items and approach the counter.

It's interesting that your brain produced this feeling for this particular number, while not doing it for the others. It only did it when the number had a meaning or some sort of significance for you. What's going on? Let's develop an experiment to find out!

Experiment: The P300 Response

The examples above are a form of decision-making. You are (perhaps subconsciously) evaluating all products while scrolling, or evaluating all numbers being called out and determining whether they match the one of interest to you. Once your brain decides there is a match, you can spring into action. What is nice about these particular scenarios is that you know precisely when your brain made the matching decision—it came soon after the information was presented. It's often hard to experiment on cognitive decision-making because it can be difficult to get a precise measurement of when decisions are made. For example, when did you decide to sit down and read this chapter? There may have been a number of factors that went into this decision. But in the case where you are waiting for a particular item, the exact time your brain makes the decision can be measured.

So let's set up a decision-making paradigm that allows us to measure brain activity while knowing precisely when decisions are made. To do this, we can make a very simple task: count how many times you hear a particular tone. We will refer to this target tone, the one you want to count, as "Boop!" To elicit a decision in the brain, we need to add a second tone, one that you should just ignore. This tone is a little higher in frequency, and we will call it "Beep!" This sound is also referred to as a "distractor," as the "Beep!" tone is distracting you from the "Boop!" tones you should be counting. So, we are all set. If we can play a tone every second, your job is to count each "Boop!" You can stop the experiment when you get to 50. We will set up the probabilities such that the ones you are counting will come very rarely, as in: Beep! Beep! Beep! Beep! Beep! Boop! Beep! Beep! Beep! Beep! Beep! In psychology, this type of experiment is called an "oddball" paradigm. Subjects are presented with a sequence of repetitive stimuli that are infrequently interrupted by a different stimulus (the oddball).

We want to see what happens when the brain recognizes the target "Boop!" sounds (the ones where we decide to take action and write down a +1 tally) compared to the distractor tones. To do this, we will record the EEG from the parietal lobe, the portion of the brain near the top of the head that is responsible for interpreting sensory information. Place the EEG headband around the chin and toward the back of the head, as shown below. The metal electrodes should be placed roughly over the P4 and Pz locations as defined in the standard EEG 10–20 system we've been using.

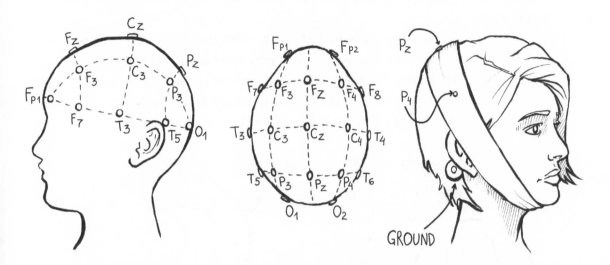

Add some globs of electrode gel underneath the metal buttons on the bottom side of the electrodes, and move as much hair as possible from between the metal and your subject's skin. Next, add a patch electrode to the mastoid process behind the ear to act as your ground, and connect all 3 electrode leads to the SpikerBox. Open the SpikeRecorder software and start recording. As with all EEG experiments, try to position your device and SpikerBox far from any electrical outlets and fluorescent lights. If you're using a laptop that has a plug, make sure it is using battery power only. If the signal seems excessively noisy and unstable, try adding more conductive gel between the headband electrodes and the scalp. If needed, you can always turn on the 50 Hz (Europe, Africa, Asia, South America, Australia) or the 60 Hz (North America) notch filters to reduce the electrical noise in software.

Have your subject sit comfortably, and make sure they are equipped with a piece of paper and a writing utensil. During the experiment, they should hold as still as they can while still being relaxed. Muscle movements from the jaw and forehead can be picked up very easily, which causes interference with your EEG reading. It is good practice to test the legitimacy of the signal before you start recording. Have the subject open and close their eyes, alternating every 10 seconds. If you can see alpha waves appear when their eyes are closed and disappear when their eyes are open, you are most likely recording a real EEG signal.

Once you've found a good signal, you can begin the oddball experiment. Start a new recording in SpikeRecorder, and press the button on the SpikerBox to start the stimuli. Once the task is started, you will begin hearing the normal (Beep) and oddball (Boop) tones every 0.5s. The tones will be randomized, with only 10% of them being a "Boop," whereas the other 90% will be a "Beep."

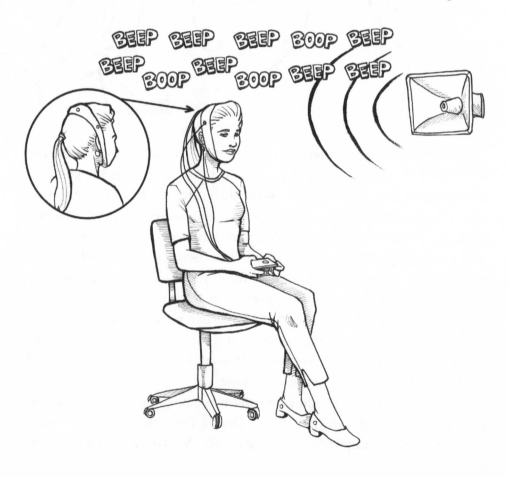

Using a pen and paper, the subject will mark each oddball tone that they hear, keeping track of the total until they've reached 50 oddball tones. Encourage them to keep focused on the task at hand, as the recognition of different tones is part of what we are studying! After 50 or so oddballs (<10 minutes), we have enough data and can end the recording.

Every time a tone was presented, the SpikeRecorder software kept track of time relative to the brain's response. Let's see what it looks like when we observe the standard "Beep" tone. This will give us a baseline for what the EEG response looks like to an auditory tone. We can begin this analysis by aligning the traces of EEG data such that the various "Beep" trials are all lined up with the tone starting at time = 0. Here we are looking at raw EEG traces from 6 trials of the standard tone. If you look through each of the trials, you will see some EEG activity, but there isn't a clear difference before or after the tone.

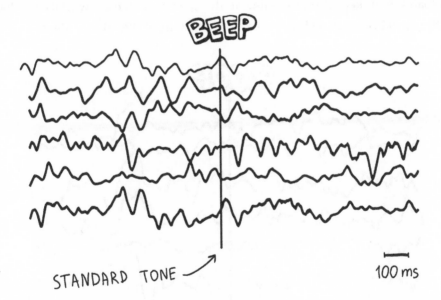

It can often be hard to spot patterns in the EEGs of individual trials, but one trick we can use is to average all the tone-aligned signals to see if there is anything interesting. The theory is that even small voltage fluctuations that are consistent across trials (meaning they tend to move in the same positive or negative direction around the same time of the trials) will be seen as a positive or negative bump in the average, while the random fluctuations (voltages that go positive and negative randomly) will average to a flat line.

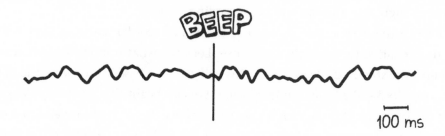

In our standard tone, there are some small fluctuations in the average of the ~450 trials, but nothing seems to stand out. So let's turn our focus to the oddball signal. Repeating the same analysis, we will take a few trials and look at the raw EEG traces. Below are six random trials of the "Boop" signal.

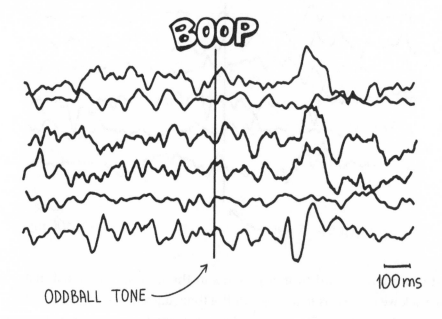

There may be something consistent happening here, but it's hard to see. Note that in four of the six trials, we can see a positive deflection after the odd-ball tone at roughly the same time. But again, to help us out, we will average the 50 trials of the oddball signal.

Here we can clearly see that those small bumps visible in the raw traces were happening often enough at the same time following the oddball tone that they stand out as a positive bump in the average.

It's always good to compare these two different stimuli on the same scale. Here we are showing the average EEG traces of both the standard and oddball tones. The gray area is what we call the "confidence interval," meaning that 95% of the EEG traces fall within the gray stripe.

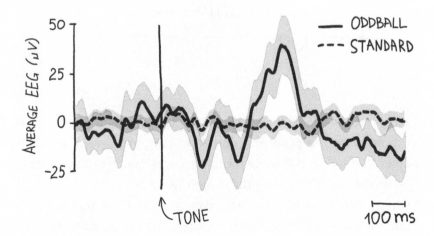

What is so fascinating about this difference is that the stimulus is relatively the same for both. They both were computer-generated tones. The difference was in your subject's mind. The subject was waiting for one of the tones to occur, and ignoring the other. This is considered to be an "endogenous" potential, meaning that the response is related not to the physical attributes of a stimulus, but rather to your reaction to it.

We call this type of signal plot an "event-related potential" (ERP). This is a different analysis than we did when detecting alpha waves. In ERPs, we average all the EEGs around a particular event, looking for voltage bumps that "lock onto" an event. The averaging makes small but consistent time-locked voltages stand out. With waves, we are looking for rhythms in the raw EEG trace after an event (e.g., eyes close). Both analyses are useful, and you would be wise to keep them in your toolbox for future experiments.

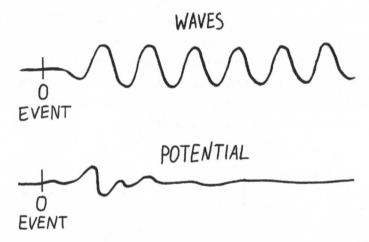

The ERP we see in the oddball stimulus is referred to as a "P300" signal. The P300 is shorthand for a positive bump 300ms after the event. Other signals have names that follow this same format, such as the "N170": a negative dip in the ERP occurring 170ms after the subject sees a face.

But why is this happening? A large part of the P300 signal is thought to come from the parietal lobe (where the signal can be most strongly recorded), which is where you positioned the headband electrodes on your subject. Many

EEG and imaging studies have postulated that the P300 arises from the brain's "attention network" that involves components of the frontal, temporal, and parietal lobes. The parietal lobe is the mapping area of our cerebral cortex, unifying our senses of touch, hearing, and sight to help us build a model of our bodies in the world. The brain's attention network is sensitive to novel stimuli and where they occur in the space around us, and that is perhaps why the P300 is strongest in the parietal lobe. Scientists continue to investigate the cause of this phenomenon.

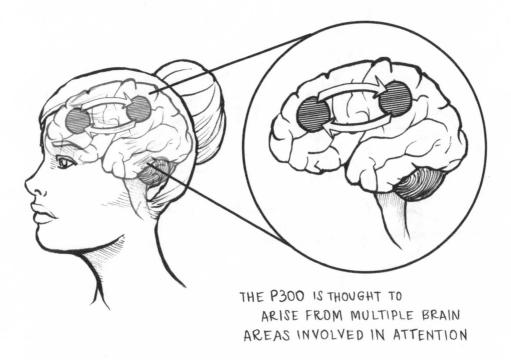

THE P300 IS THOUGHT TO ARISE FROM MULTIPLE BRAIN AREAS INVOLVED IN ATTENTION

When you see or hear something odd, or something that sticks out to you, neurons in the parietal lobe have a surge in activation as your neurons begin rapidly spiking in this area; your brain works to react to and understand this new stimulus. Again, the P300 signal doesn't come directly from stimuli itself but from your brain's assessment of the stimuli. Not all sounds initiate this signal, but rather the conscious distinction of a new sound among others.

One of the more fascinating applications of the auditory P300 is to examine the brain activity of comatose patients. Some evidence shows that if you perform this experiment on an individual in a coma and see a variant of the P300 signal in their EEG, it is a strong indicator that they might be able to be brought out of the coma. Could the P300 be a consciousness detector?

Follow-Up Questions

(1) Repeat the experiment where the "Beep" tone is being counted and only showing up 10% of the time (versus 90% "Boop"). Do the results match your hypothesis?

(2) What other stimuli could work in this way? What other ideas could work to invoke this endogenous signal?

(3) Think of other ways you can research the P300. For example, scientists have shown that the P300 brain wave tends to be large when a person recognizes a meaningful item among a list of items that are not meaningful. Could you design an experiment to test this?

(4) Since the P300 has been shown to be an attention-dependent cognitive component in wakefulness, do you think it would also be present during sleep? What about during the REM phase of sleep?

MU MOVEMENT
MIND READING

14

Mu Movement Mind Reading

Your muscles power the machine that is your body. Carefully applying the appropriate force, they contract in just the right direction with incredible timing and coordination with other muscles. Your brain can effortlessly control specific muscles required for a task (like picking up a heavy bowling ball), while subconsciously controlling other postural muscles to maintain your balance. The responsibility for all this muscle control is relegated to your motor cortex, and it is truly your brain-machine interface. In fact, the only way the brain can interact with the physical world is via the motor cortex. The brain commands and the motor cortex does its bidding, allowing you to talk to others, shake hands, hug, and fight. It's hard to overstate the importance of our motor cortex in the human experience.

Even the word "emotion" derives from the Latin word *emovere*, meaning to move, agitate, or stir up. We talk about "being moved" by beautiful art. Emotions are not just a feeling inside your head; they also manipulate (via the motor cortex) the physical ways we express these emotions through facial expressions, voice pitch and volume changes, and our body posture. Because these emotional circuits access the same motor cortex we use for our own movements, it can be nearly impossible to hide the way you are feeling. Through millions of years of evolution, the brain has learned what motor patterns are best suited for particular emotional situations. Patterns that were considered useful for the emotional state slowly turned into reflexes, then into instincts, which can be hard to override. We frown when concerned, smile when happy, close our eyes when struggling to open a can of pickles. We can consciously override these expressions through our motor cortex, but as soon as we are distracted, the true subconscious expressions will take back the reins of our motor cortex.

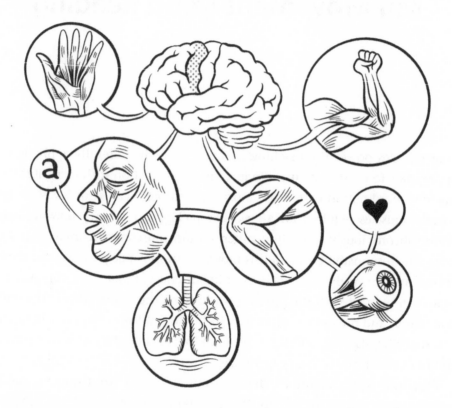

The motor cortex is also your last checkpoint for decision-making. Once a spike command leaves the primary motor cortex, it sets off a chain reaction of neurons that will end up with a movement. There is no going back. So when deciding to swipe left or right, it's your motor cortex that signals your final decision. This makes the motor cortex an ideal spot to place electrodes to have the brain control external machines in the real world (for example, prosthetic limbs or cursors on the screen). Not only can it encode motor sequences, but it also will only fire once you have made the decision to move, avoiding unintentional or unplanned movements.

In this chapter, we will look at recording from the motor cortex. Not by implanting electrodes into your head to record the beautiful spike trains that give rise to movements, but by recording safely from the outside of your head using electroencephalograms.

Experiment: Mu Rhythms of the Motor Cortex

Let's begin exploring the motor cortex using our EEG headband with rivets. Place your sweatband electrode across your head as if you're about to play tennis, then position the metal leads to the side so that they cross above your right ear. This location is ideal as the motor cortex is a vertical strip that runs from the top of your head to the front side of your ears. These placements are called C4 and F4 in the 10–20 EEG system. You may want to become familiar with using these standard location names, as they will allow you to compare your results with other scientists.

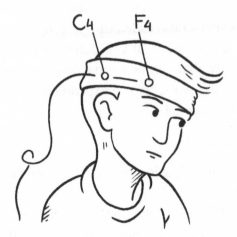

Once the electrodes are in place and you have enough gel under the metal buttons, connect the ground behind your ear and plug in the electrode cable to your SpikerBox. Find a place to sit and rest, then turn on the device and watch the recordings. Be still and observe what types of signals you see.

1 s

If you are above the motor cortex, you might see a few different patterns. But the up/down waviness of the EEG is often quite dominant in this motor area. These ripples, or oscillations, can ring within the EEG with a slight ebb and flow in their size, as shown above. We can describe this signal using the frequency of the dominant oscillations. Using our "back of the envelope" calculation, we can count the number of peaks or the number of valleys within 1s. In the trace shown above, you can count the number of downward valleys above the 1s time bar to be around 12. Congratulations—you just captured the mu waves! This signal (also called the "mu rhythm") is observed at 7.5–12.5 Hz above the motor cortex. If you don't see these synchronous waves above the motor cortex, that may be normal. Strong mu rhythms are not seen in everyone. If it doesn't work, try the experiment with another subject. Be sure to add lots of electrode gel.

Experiment: Desynchronization of Mu Rhythms during Movements

Now that we found the mu rhythm, let's see what it does when our motor cortex causes movements. Relax your right hand for 10 seconds, and then squeeze your hand for another 10 seconds. Be sure you (or your assistant) mark the beginning and end of movements in your recording. Now go back and look at the recordings. What did you notice when your right hand closed and opened? Nothing much?

Now do the same thing with the other hand. While still recording from the right hemisphere of the brain, open and close your left hand. What do you notice now?

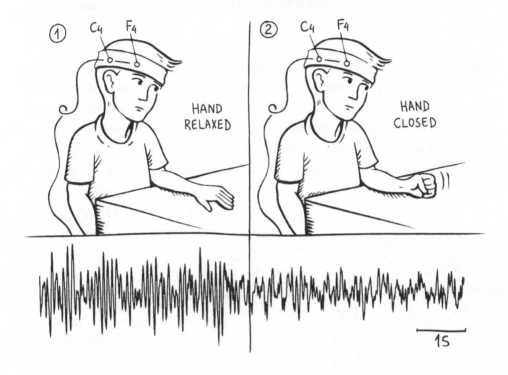

The mu rhythms tend to go away or get smaller when you control the hand opposite from where you are recording on the brain. We call a rhythm that disappears in the EEG as being "desynchronized." To understand why, remember how the EEG rhythms show up in the first place. Many neurons need to be firing a small voltage in the same place at the same time—in other words, they must be synchronized. If they don't, they become desynchronized, and the small voltages will cancel each other out, leaving a weakened or absent signal in the EEG. OK . . . but why does this happen on the opposite side of the brain? This is due to an odd quirk in our bodies' wiring. The axons that exit the motor cortex don't go straight down to the spinal cord as you might expect, but instead cross over to the opposite side of the spinal cord via the pyramidal tract. Why do the motor signals cross over? No one knows for sure.

Experiment: Detecting Mu Rhythms Desynchronization from Different Body Parts

The synchronized mu rhythms can be across the entire strip of the motor cortex from ear to ear. By moving the headband to different positions of your scalp (keeping centered on the motor cortex), see if you can map out your homunculus by voluntarily moving different parts of your body and observing mu desynchronization. Try recording toward the top and bottom of the motor cortex. Can you see a difference when you move your tongue? How about your leg?

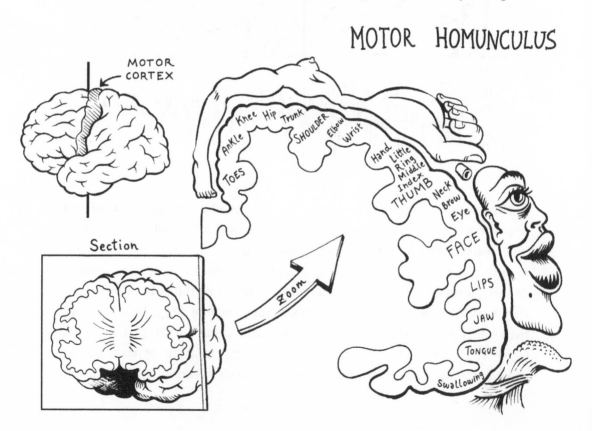

Experiment: Differentiate Alpha Waves from Mu Rhythms

We have seen very similar waves to these mu rhythms before. In our initial EEG chapter, we saw a ~10 Hz frequency called the "alpha wave." Are we sure these two rhythms are not the same phenomenon? Let's recall that alpha waves are most prominent in the brain's visual cortex. They have a higher power when a person has their eyes closed, blocking off the visual input. Mu rhythms, on the other hand, were seen above the sensorimotor cortex when a person is at rest. Let's confirm that these two signals are indeed separable.

Plug your SpikerBox into your computer and open up the recording software. This time, you will use two different signals: one across the motor cortex as before, and one in the back over your visual cortex. Again, place an electrode sticker behind your ear (mastoid bone) to be your common ground for your recording. Be sure to apply plenty of electrode gel under the headband electrodes to ensure you have a strong, conductive contact to the scalp.

In this experiment, start off by closing your eyes. Wait a few seconds, then open your eyes. Wait a few seconds more and close your fist. With your eyes still open, relax your hand. Close your eyes and repeat this process. Do this for a few minutes, while your partner marks your eyes open/closed and hand open/closed in the session recording.

Once you're done, open the recording file and observe as it plays back. Every eyes-closed section should show distinct waves or oscillations in the visual area. These are our old friends the alpha waves from chapter 10, which originate from the visual cortex. You may even see this rhythm "spill over" into the sensorimotor cortex.

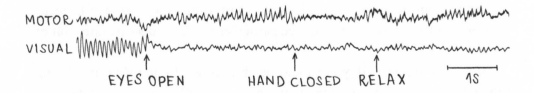

However, the segments where you were relaxing with your eyes open may show waves that resemble alpha over the motor cortex. They are still noticeably rhythmic, only smaller and a bit transient—meaning they appear and disappear more quickly. These could be the mu waves! To verify, look at what happens when the hand closes. Note in this experiment that the two waves have nearly the same frequency, but it is the location on the brain that differentiates the two. When looking for mu rhythms, it often helps to observe both the visual and motor areas to be able to tease the two signals apart.

This desynchronization in the mu waves may be the result of your sensorimotor cortex leaving the "idling" mode, preparing for and initiating movement. This is a similar theory to the idling alpha rhythms when there is no visual input. If this is true, even imagining a movement should have the sensorimotor cortex ready to leave this idling loop, as much as an actual movement.

Experiment: Mu Rhythms Desynchronization in Imagined Movements

To test that hypothesis, perform the experiment one last time. Only this time, instead of moving your hand, simply imagine moving your hand. Do about 10 seconds of intensely thinking about and visualizing forming a tight fist with your hand. Try not to flinch or even make small movements. Imagining does not, however, mean just a visual imagination—it is more like focusing on thinking about moving your hand but never actually moving it.

In the relaxed state, imagine not moving your hand. Nothing should be changing from an outside observer's perspective throughout this experiment, so this can be hard for the test subject. When someone asks you not to think about movement, chances are that's the only thing you'll be able to focus on. So, managing to willingly dodge particular thoughts does require a bit of training.

If everything worked, you should be able to see the mu rhythms during the relaxation periods, and then you can see them disappear during the periods of visualizing that physical movement. If you had a series of electrodes along your scalp, could someone reading only your EEG see what body part you are imagining moving? Try it out! See if you can make a movement mind reader!

In fact, a lot of research has been conducted on mu rhythms, and they continue to be explored for the development of brain-computer interfaces (BCI) in the development of neuroprosthetics. Devices that scan for desynchronization events in mu rhythms can be used to control simple video games. While the software used for analyzing and decoding these signals are a bit more advanced than we can do by eye, they follow these same principles. Could the mu waves be a strong contender for future, easy-to-use BCI devices? Maybe you can help decide.

Follow-Up Questions

(1) Repeat the experiments above, but instead of the subject moving their body, have them look at someone else doing the movements. Some researchers have found that mu suppression in observing other movements may be an index of mirror neuron activity.

(2) Some research suggests that the visual recognition of easily graspable objects causes quick and strong desynchronization, priming the sensorimotor cortex to perform that action. Imagine recording someone's mu rhythm, and a tennis ball or apple is placed on the table within arm's reach. Would we see the mu rhythm change? Now use a large object we don't normally grasp with one hand—a large flower vase filled with water and flowers. Would that change the mu rhythm?

YOUR BRAIN ON

meditation

15

Your Brain on Meditation

The world is moving toward a broader awareness of wellness. Better health means not just eating right and maintaining good physical exercise (often carefully tracked by fitness devices), but also doing mindfulness exercises such as yoga or meditation. We are awash with advertisements for apps and videos that promise to help train your brain for you to reach a more peaceful and calming state and improve your mental and physical health.

Such lofty promises do seem to be based on data in some regards. The effects of mindfulness meditation have been researched and published in medical journals, and indeed many studies have shown its benefits in reducing blood pressure and stress. But unlike diet and exercise, which are easily measurable and quantifiable (calories eaten, miles walked), mindfulness meditation relies on qualitative self-reporting. This gives fuel to skeptics and could make it difficult for beginners to learn. But mediation should be having a physiological effect on your brain, given the reports of meditation experiences. And these effects should be different inside your brain than when you are simply resting quietly with your eyes closed (although an outside observer would not be able to tell). If we could record from the brain during rest and during meditation, we may be able to tease apart these differences and help newbies by adding a unit of measure to meditation.

Experiment: EEG of Meditation and Rest

For this experiment, we are going to use our SpikerBox to record EEG data from a mindfulness meditator. Find some friends or relatives who are good at meditation . . . those who tell you that meditation really *feels* like something. As controls, you can also recruit others who do not meditate. Once you find someone willing, have them sit still as we prepare to record from four scalp locations. This is different from the other EEG experiments in this book, which typically only use one or two locations. The reason for adding more is that we do not yet know where to look for any changes. So we are going to expand our search area and make our hypothesis that meditation should change brain regions associated with visual, auditory, executive, or touch sensory integration. This makes sense as we have seen that the lack of input (both visual and motor) gives rise to oscillations in the EEG. Perhaps we will find something similar when our subject "clears their mind" during meditation. To get started, you will need two sweatbands with metal rivets, and arrange them so the metal electrodes line up with the locations shown in the diagram below.

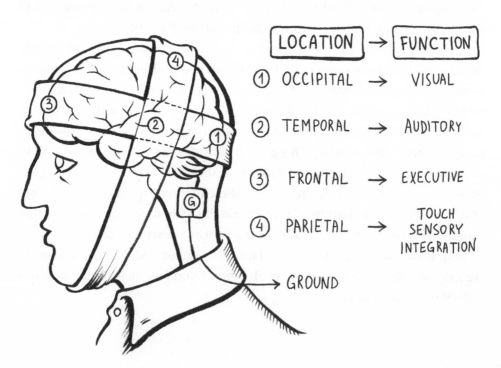

These recording locations correspond to different lobes of the human brain that have been identified as hubs of certain functions through research studies. The occipital lobe at the back of the brain accounts for visual functions. The temporal lobe deals with auditory stimuli, taste, face recognition, and memory. At the front lies the frontal lobe (you guessed it!), suitably entrusted with the executive function. Finally, the parietal lobe is in charge of touch and sensory integration. Don't forget your ground electrode behind the ear! Note that you can clip all the ground alligator clips together on this one patch electrode.

Plug your cables into your SpikerBox and open up the SpikeRecorder software. Once your device pops up, connect it and start a new recording.

Rest versus Meditation . . . Fight!

We are going to now attempt to distinguish between a brain at rest and a brain during meditation. But as good experimentalists, we will try our best to remove any strange effects that may result from external factors that could interfere with our results. One thing we should be careful of is the order in which we have our subject rest versus meditate. If you always kept the same order—say, rest then meditation—you could end up fooling yourself that there were rest/meditation differences where there weren't. Say, for example, the room was hot, people were getting tired toward the end of the experiments, and you could end up claiming that the sleepy brain activity was actually a meditative state! To solve this problem, you can simply flip a coin. If it is heads, your subject will begin with a 10-minute period of rest. If it comes up tails, pick the alternative—10 minutes of meditation. The independent randomness of the coin toss ensures that any difference between the rest state and meditation state is not based on ordering.

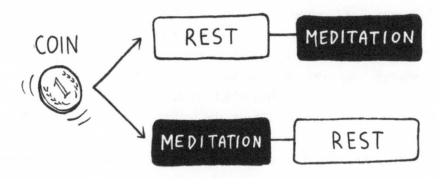

In rest, the baseline task, your subject will sit quietly with their eyes closed for 10 minutes without meditating. During meditation, your subject will perform their regular meditation technique. They may focus on their breathing, allowing their body to calm itself. They may fixate their mind on a single object or notion while trying to turn off or cast aside every possible distraction. If mindfulness is not your subject's cup of tea, during the meditation sessions, ask them to attempt to let go of all control, allowing their train of thoughts to flow freely . . . Judge not thy thoughts lest ye be judged! Mark down in each session the person's ability to meditate and the technique used. Be sure to ask how well each session went, as we will need this later to compare. Whichever technique is used, have the subject meditate or rest in 10m blocks. If you can get someone for an hour session, you can have three rest periods and three meditative periods. Having multiple sessions from the same subject is useful to help determine whether changes are consistent or perhaps by chance.

Analysis

Once you have collected data from several subjects, it's time to start looking at the data. The first task of data processing is to remove any artifacts, meaning periods of time where the headband wasn't making a good connection (creating periods of noisy signals) or even other biological signals that may appear as EEG. For example, when recording the frontal area, you may pick up the muscles of the eyes during eye blinks. We can see in the example below that eye blinks can show up as small deflections in the data, and may convince you that you are recording delta waves. We will want to look at large periods of time that are free from such artifacts.

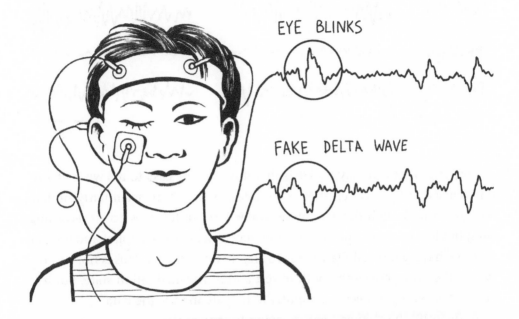

EYE BLINKS

FAKE DELTA WAVE

A good way to start analysis is by simply scrolling through your EEG data on your screen and jotting down any noticeable differences between the states of meditation and rest. For example, you may see something like this for a self-reported expert meditator (someone who mediates more than 5 days per week).

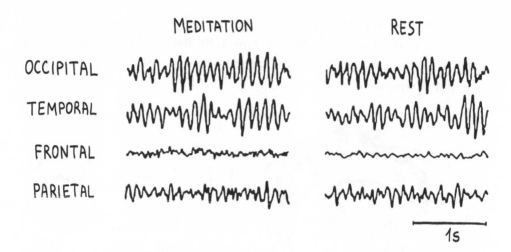

The first thing you may notice is that it is not very easy to tell what, if any, difference there is between these two states, as they both seem fairly similar. As you scroll through the data more, you will see differences across recording sites, but less so across conditions. You may observe the occipital and temporal lobes having more alpha waves (10 Hz) then the others. This is not surprising, as the eyes were closed both in rest and in the meditation state, but it is clear we are going to need a computer to help us analyze the data.

To quantify the changes, we are going to look at the power in specific frequency ranges of the EEG. You can visually observe the power by looking at the amplitude (the height from valley to peak) of the oscillations. The bigger the peaks are, the easier they are seen from the noise, and the more power they contain. Historically, the various EEG oscillations are often divided into specific frequency classes that are commonly seen in recordings. While the exact frequencies of each class often vary across papers from different research groups, they are typically found as follows:

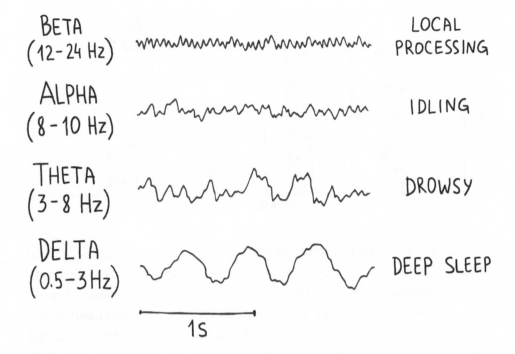

BETA
(12-24 Hz)   LOCAL PROCESSING

ALPHA
(8-10 Hz)   IDLING

THETA
(3-8 Hz)   DROWSY

DELTA
(0.5-3 Hz)   DEEP SLEEP

1s

For our analysis, we are going to focus on changes in the power of alpha frequencies between the two conditions. Why alpha? This comes from intuition (the brain should be idling if thoughts are emptied during meditation), as well as from reading research papers from the EEG literature. Other studies have shown meditation differences in the alpha range. Using a computer, we can determine the average power of the alpha band in both the 10m rest period and the 10m meditation period. Here is what this analysis looks like for our expert meditator:

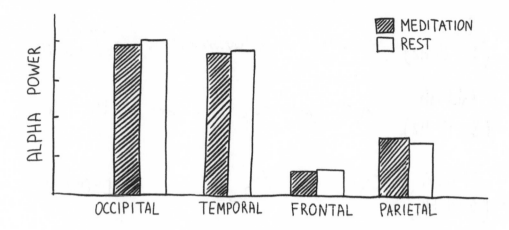

Hmmm. There does not seem to be much of a difference in any of the four regions we chose for this study. Even with careful frequency analysis on a computer, this meditator seems to have similar size alpha waves no matter if at rest or in deep meditation. In fact, you may find that there were not many differences across any of your subjects! And if you did, these differences may get lost when you group together sessions from all your experienced mediators. Grouping together multiple subjects is the best way to make claims about a phenomenon. The effect may be small in an individual subject, but if a group of people experience the same small effect, it becomes easier to see. That was not the case here.

So does this mean that no changes occur in the brain's EEG when someone is meditating? Not exactly! This experiment goes to show that often science isn't hard due to complex formulas and jaw-breaking Latin denominations. It's hard because you have to ask the right questions and choose the right approach. We were careful to control the order of the trials and to remove noisy data . . . but the meditation approach was left up to each person. Studies that have shown effects tended to come from particular groups of meditators who tend to have similar styles (say, traditional Tibetan Buddhist meditation).

Follow-Up Questions

(1) Besides choosing people who utilize the same meditation styles, what other changes could be made to the experiment or to the analysis to better explore any differences between meditation and rest?

(2) We chose to record from four EEG locations. That's more than any other experiment in this book. But is that enough for this study? How would you change the electrode positions or number?

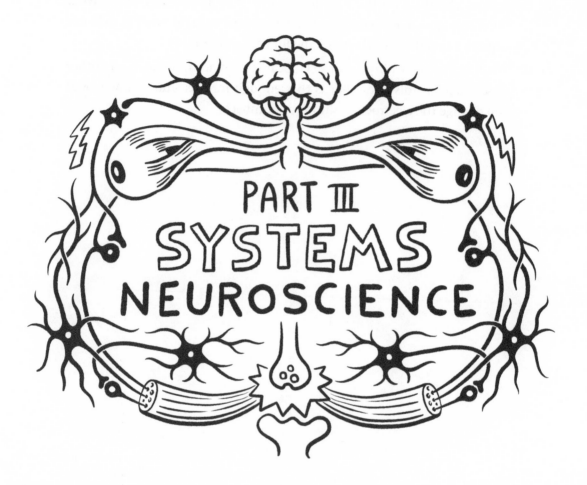

PART III
SYSTEMS
NEUROSCIENCE

III
Systems Neuroscience

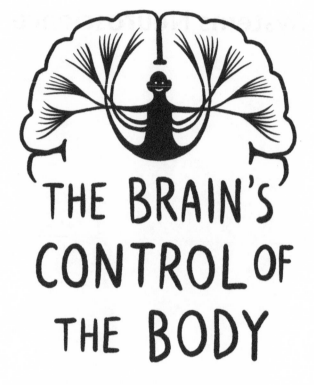

THE BRAIN'S CONTROL OF THE BODY

16

The Brain's Control of the Body

So far we've explored how recording from individual neurons (spikes) and from large groups of neurons (evoked potentials and EEG) allowed us to gain a deep understanding of how parts of the brain behave when exposed to various stimuli or controlled movements. In this section, we can start to focus on how neurons in these various brain regions interact to produce certain behaviors. To do this, we need to explore the electrical signals produced by a few other organs in the human body, starting with the heart.

Compared to the brain, the heart does a relatively boring job. Pump, pump, pump. It works continuously, day in and day out, to supply blood and oxygen through your body. Anatomically, our hearts are made up of four chambers, of which the upper two chambers are called the atria and the lower two are called ventricles. The muscles of the heart work together to fill and empty the chambers in a particular order to allow the blood to flow through your entire body. To understand how these chambers synchronize and work together, let's take a deeper look using our electrophysiology skills.

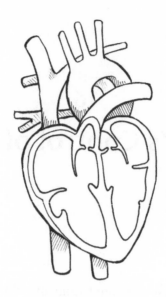

Experiment: Measuring Heart Action Potentials (EKG/ECG)

We are going to monitor our heart rate and examine changes in it caused by our brain's cardiovascular center. As you are now accustomed to, you will be using electrode stickers to record. These stickers have a circular glob of salt-water gel on one side that is surrounded by a sticky tape. On the other side, there is a metal clip that touches the gel and allows our electrode leads to connect, providing a good connection to the body. Small electrical currents can easily travel from the body to the skin, through the saltwater gel to the metal clip, then down our electrode cables leading to our SpikerBox. There are many options for where we can place our electrodes.

One of the easiest ways to record the heart's signal is by placing the patch electrodes on the insides of each of your wrists, with a ground on the back of your hand. Clip the red alligator clips to the patches on the insides of your wrists and the black alligator clip to the ground on the back of your hand. This signal can get a little noisy though; any wrist flexion can muddy the signal and make it difficult to interpret, so be sure to rest your hands on a table and relax. If you are having trouble picking up a signal, there is also an option for a chest reading. This is a bit more difficult due to the barrier of your clothes, but it produces a much larger signal that is less affected by noise. Place the recording patch electrodes across your heart on your upper chest, and another patch for your ground on your lower chest. Connect the red alligator clips to the heart electrode patches and the black alligator clip to the ground electrode patch.

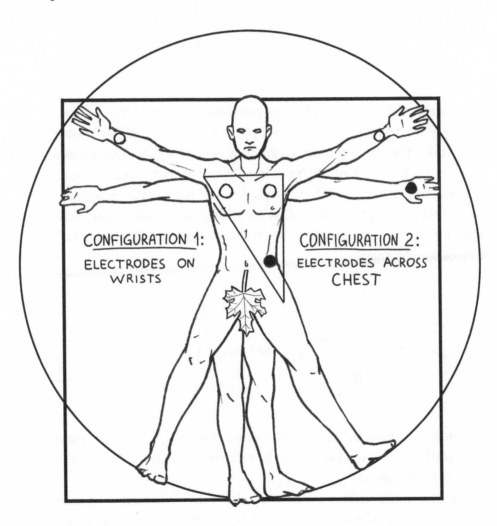

CONFIGURATION 1:

ELECTRODES ON WRISTS

CONFIGURATION 2:

ELECTRODES ACROSS CHEST

Connect your electrode cable to your SpikerBox. Fire up your computer or smartphone, and let's start recording! At first, your signal may look different from what you might be used to from visits to hospitals or from TV shows:

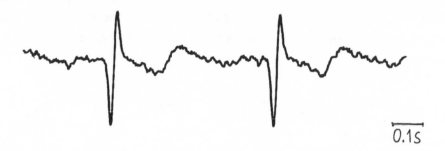

0.1s

Something is not right. The first peaks are pointing down instead of up! To figure out what is wrong, let's remember how a bioamplifier in the Spiker-Box works. The signal we are recording is an amplification of the difference between the two red leads (let's call them A and B). With a gain of G, this simply turns out to be G(A–B). But what happens if the leads are reversed? We get G(B-A), which is equivalent to –G(A–B). The signal is upside down! To fix this, simply swap the locations of the two red leads, and your EKG should return to a more familiar shape:

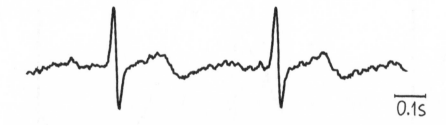

0.1s

Ahh . . . much better! Behold the electrocardiogram! This is also called ECG or EKG (these abbreviations mean the same thing . . . only the German spelling is *elektrokardiogram*).

If your signal is noisy, try a different electrode configuration or move your recording device to a different part of the room. If you are using a laptop, avoid plugging it into the wall for power as the alternating current (50 or 60 Hz) from the outlet can introduce noise. Also, even small muscle movements can be picked up, which can cause interference with your EKG reading. Trying to rest with your hands on your knees provides the most stable signal.

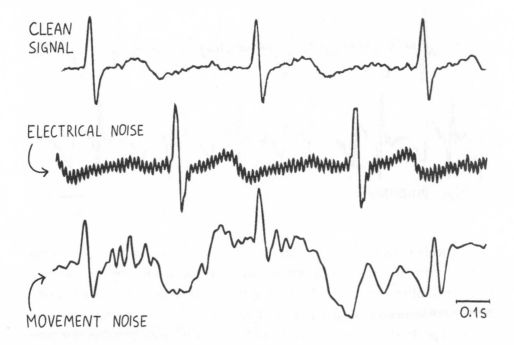

CLEAN SIGNAL

ELECTRICAL NOISE

MOVEMENT NOISE

0.1s

OK. So this looks familiar. But how do we know this really is your heart signal? Try to gently feel your pulse while watching the EKG. Do you notice the correlation between the mechanical signal (pulse) and the electrical signal you see on the screen? Each pulse should be associated with one peak in the signal, suggesting that this signal may indeed be coming from the heart. Let's try one more experiment to be sure.

Experiment: Electrocardiographic Response to Exercise

Let's test this relationship by changing the speed with which you pump blood. Disconnect your electrodes and move your body a little bit. Do some jumping jacks, jump rope, or run in place for about 30s to increase your heart rate. When you reconnect your electrode cables, you should see your increased heart rate on the screen.

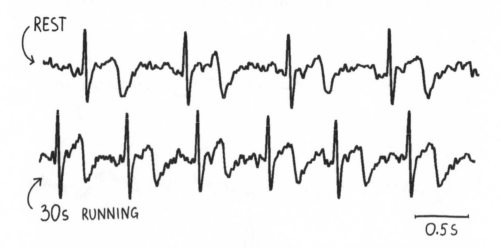

To calculate the number of heart beats per minute, a standard unit for discussing heart rate, we can take one minute (60s) and divide it by time between two consecutive peaks. For the resting heart rate above, the time between beats is 0.93s, or 65 beats per minute (BPM; 60s/0.93s). After exercise, the time between peaks shrinks to 0.6s or 100 BPM. This increase matches our intuition, and suggests that the electrical signal is definitely related to the heart.

But what is this electrical heart signal, and why can you pick it up so clearly even on your inner wrists? Let's take a look at the heart physiology. The heart contains special myocardial cells, called "pacemaker cells," that spontaneously generate rhythmic action potentials to control your heart. These pacemaker cells rhythmically cause the rest of the heart's myocardial cells to contract in a specific pattern (atria first, delay, then ventricles). This is why your heart can beat on its own without neural input.

These pacemaker cells are vitally important in telling the rest of the heart to contract in a synchronized manner. (When the cardiac muscle cells fire and contract in a disorganized manner, this is called "fibrillation" and can be fatal.) Imagine a crowd doing "the wave" in a stadium sports event. Many people have to be precisely coordinated for the wave to occur, and similar principles apply to heart contraction. Select groups of cardiac muscle cells must contract at the right time for effective pumping.

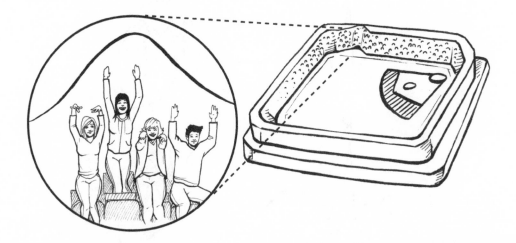

OK. But how can we record a signal from the heart on the wrists? It's similar to why we can see the EEG waves through the scalp. This happens when the brain's electrical activity is coordinated by a large number of neurons firing in the same place at same time. The same is true for the heart. The coordinated contraction of closely packed heart muscle cells allow the electrical signal to travel a great distance in the body (including your wrists)!

Now, let's take a closer look at these contractions. The normal heart cycle has a very distinctive electrical profile: the classic P wave, QRS complex, and T wave.

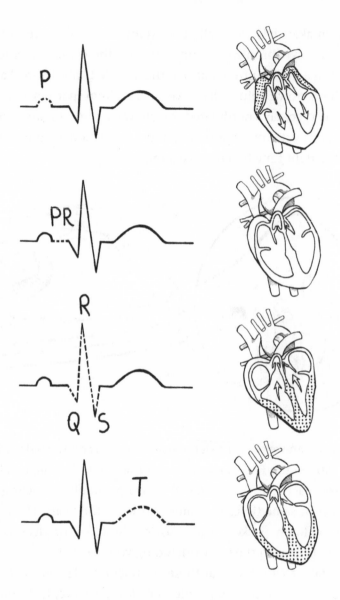

These electrical signals are regulated by your brain's cardiovascular center located in the medulla oblongata. The medulla is the last stop in your brain before the spinal cord, and it controls basic bodily functions such as breathing, heart rate, and blood pressure. Interestingly, your heart can beat without instructions from the brain. But many times, your brain takes the reins and decides your heart rate needs to be adjusted. During exercise, it's the cardiovascular center in the lower brain that sends signals to your heart, changing both the rate of your heart's beating and the strength of its contraction.

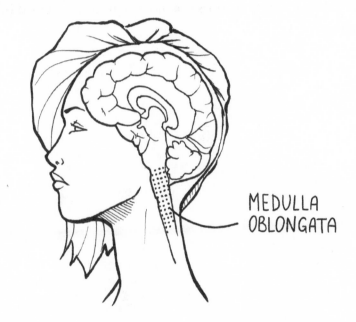

MEDULLA
OBLONGATA

Now that we can easily measure the EKG, we can begin to take a deeper look at how the brain can affect control of the body.

Follow-Up Questions

(1) Are we recording cardiac action potentials from the pacemaker cells, or are we recording the muscle contractions of the heart? Why?

(2) There are lots of different places you can put the electrodes for this experiment. What effect(s) do the different placements have on the signal you see? Is the signal different if you move the electrodes closer to one another? Closer to the heart? Why?

(3) Do you think your resting heart rate is sensitive to outside temperature? Try comparing your resting heart rate on cold days versus hot days. Stay safe though!

(4) Would your resting heart rate change before and after eating? Why might that be? What, if anything, does the movement of your blood have to do with eating?

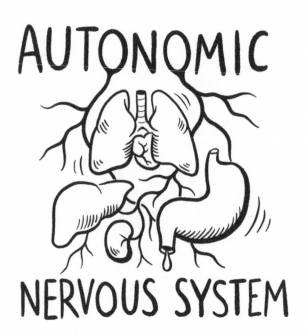

AUTONOMIC NERVOUS SYSTEM

17

Autonomic Nervous System

Our nervous system is not just centralized in our heads. It runs through the entire body. But why does the nervous system need to spread around the body? Let's think about the last time you had a big exam. Before taking the exam, you probably had butterflies in your stomach. You may have even felt your heart racing, your skin sweating, and your head pounding. It seems like the mere thought of the exam in your brain is enough to prod the rest of your body to prepare for action.

Now think about how you felt after having a large meal over the holidays. You probably felt tired and relaxed. In fact, your body may have started digesting the food before you even finished eating. It seems your body is lulling you into a state of rest and sleep. In this chapter, we will perform some experiments to examine how the brain can control the rest of the body to best prepare for what lies ahead.

Experiment: Sympathetic Nervous System Activation

In this experiment, we will need to concoct a stressful situation so we can carefully watch how the body reacts. But the hard part is figuring out how to apply stress that is *real* (it's hard to be stressed over an exam that doesn't really exist) and *ethical* (we don't want to feel physically threatened). Fortunately, there is a simple way to make someone stressed, without resulting in psychological damage: ice water! A bucket of ice water is often used in pain studies as humans can tolerate it, everyone has experienced cold hands before, and it is not scary. Also, it is a good model stimulus that is easy to replicate in labs around the world. The longer you keep your hand in ice water, the more uncomfortable it becomes. Let's see what happens to our EKG when we are beginning to experience this sort of stress.

Fill a large receptacle, like a large bucket, three-quarters full with ice, then pour in cold water. This ratio is important, ensuring that the mixture is always in equilibrium at 32°F (0°C). You will be placing your arm into the bucket, so be sure to place the patch electrodes high enough on your upper forearms (near your elbows) so the red alligator clips won't get wet. The black alligator clip can attach to a ground electrode placed on the back of your non-submerged hand. Next, plug your electrode cable into your SpikerBox and connect it to your computer or recording device.

Submerge your hand in the ice water, leaving your upper forearms exposed—be sure to keep the electrodes dry! Leave your hand in the ice water until you can barely tolerate the cold any longer, and write down your heart rate at that moment. Let your hand relax and warm up for a while. Repeat this process a number of times so that you can get an average.

Experiment: Parasympathetic Nervous System Activation

We will repeat the experiment protocol above, only this time you will dip your entire face into cold water instead of your hand. We don't want ice cold water this time! Dump out your ice water and replace it with more tolerable cold/cool water. Make sure the container is big enough to dip your face . . . and keep your own safety in mind! Always have a buddy supervising you during this experiment.

To begin, record your resting heart rate as your baseline. Next, hold your breath and submerge your face underwater for as long as you can do it comfortably. Have a partner observe the heart rate after your face touches the water. Note any change and what is happening when the change occurs. Does it decline, and then increase when you remove your head from the tub of water and begin breathing again? Repeat the experiment a few times to get an average heart rate for both conditions.

What you are experiencing is called the "diving reflex." When a sea lion or other marine mammal dives, its heart rate decreases, and the veins and arteries in peripheral tissues and limbs contract. This limits blood flow to organs not related to the dive, reduces oxygen consumption of the heart, and maintains blood flow to the brain.

But it turns out that this response exists in all mammals, including you! When cold water contacts your face and you are holding your breath, we can see the diving reflex as a decrease in heart rate. We can try variations to separate out the effects of water contact on the face versus simply holding your breath.

So, what is happening? Most of our human physiology experiments so far have dealt with the voluntary nervous system (neuromuscular neuroscience) or perception (sensory neuroscience), but now we are focused on the "involuntary" part of the nervous system: the autonomic nervous system. The autonomic nervous system controls things we are both aware and unaware of but generally do not have much control over: digestion, homeostasis, perspiration, blood pressure, heart rate, and many others. It is traditionally divided into two systems: the sympathetic division (which activates the coined "fight or flight" response), and the parasympathetic division (which activates the also-coined "rest and digest" response).

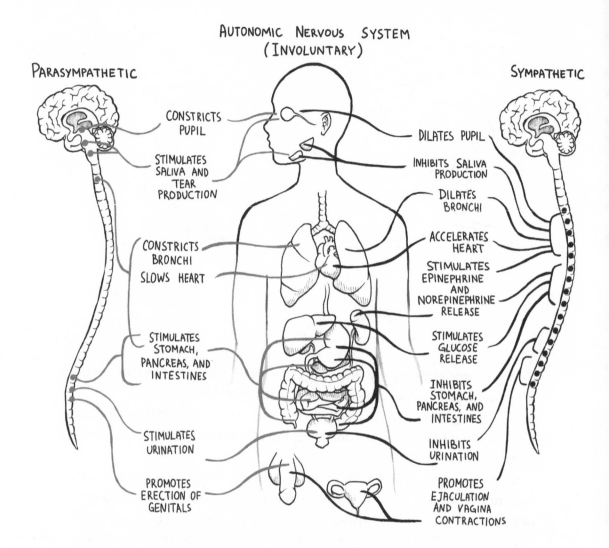

AUTONOMIC NERVOUS SYSTEM
(INVOLUNTARY)

PARASYMPATHETIC

SYMPATHETIC

CONSTRICTS PUPIL

STIMULATES SALIVA AND TEAR PRODUCTION

CONSTRICTS BRONCHI

SLOWS HEART

STIMULATES STOMACH, PANCREAS, AND INTESTINES

STIMULATES URINATION

PROMOTES ERECTION OF GENITALS

DILATES PUPIL

INHIBITS SALIVA PRODUCTION

DILATES BRONCHI

ACCELERATES HEART

STIMULATES EPINEPHRINE AND NOREPINEPHRINE RELEASE

STIMULATES GLUCOSE RELEASE

INHIBITS STOMACH, PANCREAS, AND INTESTINES

INHIBITS URINATION

PROMOTES EJACULATION AND VAGINA CONTRACTIONS

Using heart rate as an indicator, we can study sympathetic and parasympathetic nervous activation, but the effects are seen all over the body. The ice water experiment activated the sympathetic nervous system, hence the increased heart rate. The diving response causes a lower heart rate, indicating an increase in parasympathetic activity to the heart.

Many of the body's reactions in both sympathetic and parasympathetic systems are controlled by hormones, which can be helpful to think of as neurotransmitters that enter the bloodstream instead of the synaptic cleft to find their targets. This means that instead of response times of 1ms in the brain, hormones have response times in the scale of seconds to minutes on multiple structures in the body.

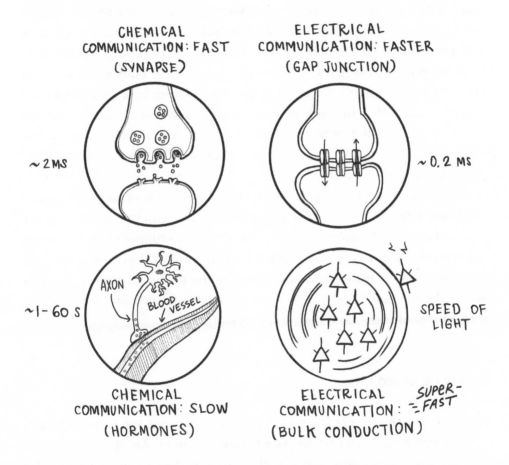

For example, when the sympathetic nervous system is activated, the pituitary gland, which anatomically branches off of the hypothalamus in the brain, releases an adrenocorticotropic hormone (ACTH) into the bloodstream and increases cortisol levels, causing various physiological changes, including heart rate increase. Simultaneously, the adrenal gland, a neural ganglion located on the kidneys, releases norepinephrine and has a similar effect on the heart.

Why is this important? Because it helps us survive! Regulating your body's blood pressure, oxygen, temperature, and breathing is an important (but boring) job for the brain to do. These housekeeping functions are relegated to the oldest part of our brain (the brainstem), which takes care of keeping important systems in a manageable range automatically. But during evolution, we needed a way to modify the brainstem's control of the body so we could be prepared to run or fight, if needed. This is now handled by a set of deep brain structures called the "limbic system" which evolved to automatically control the brain stem during certain emotional situations. This autonomic system allows us to more easily escape from a burning building or a hungry lion by priming your body for action.

Our neocortex was the last structure to evolve and it in turn can control the limbic system (which in turn controls the brainstem) in response to complex and abstract situations. For example, if there is a runner on third in baseball with two outs in the bottom of the ninth, you will feel your autonomic nervous system at work (no matter if you're the pitcher or batter). So think about this the next time you get nervous and notice you are sweating—it's your autonomic nervous system doing its job!

Follow-Up Questions

(1) Since these experiments are relatively easy to do quickly, you can rapidly generate a large data set in your family or school. Are there differences between athletes and people with normal or low levels of fitness? Are there differences across age? Differences between male and female students? Happy statistics!

(2) We previously studied the effect of exercise on heart rate. Would this stress response caused by ice increase the heart rate through different physiological mechanisms than exercise?

(3) Try using a snorkel to breathe while your face is under the water. What happens now? Does your heart rate change in the same way as when you were holding your breath underwater? What could be happening?

(4) Is the control from cortex to limbic system to brainstem only in one direction? Can you think of ways that the limbic system can control the cortex? (Hint: think of decisions you've made while under stress or in love.)

EMG
MOTOR
MOVEMENTS

18

Motor Movements

Moving allows you to explore and interact with the world. You walk, you run, you dance, you sing. You jump and you turn. But how does your brain tell your body what to do? We've already seen that a particular strip of the brain called the motor cortex (stretching roughly from ear to ear) had EEG oscillations that would disappear during movements. From there, we argued that the neurons in the motor cortex were active and sending movement commands around the body. But what is happening at the receiving end of that signal? How does the brain actually make us move? Let's hook up some electrodes and see what we can discover!

Experiment: Recording Electromyograms (EMG)

To investigate what is happening during muscle movements, let's connect our electrodes on a muscle that is easy to control and record from. If you wanted to record from your arm, place two patch electrodes close to each other on your bicep. Always try to spread the stickers apart but keep them over the same muscle to help isolate the signal in the particular muscle you are recording from. You'll want to connect the red recording alligator clips from your electrode cable to these two patch electrodes. Next, connect the black ground alligator clip to an electrode patch on the back of your hand or perhaps any metal jewelry you are wearing.

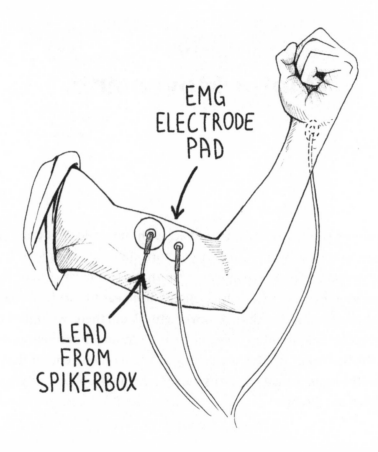

Plug the electrode cable into the SpikerBox and turn it on. Start by listening to the sounds that occur. Try to detect changes in the sound—for instance, when you flex your bicep. You will hear a similar sounding popping noise of spike activity as you did for our invertebrate neuron experiments, only this time they sound a bit deeper, or muffled. Flex a few times, slowly and quickly, to convince yourself that you are indeed recording something from your own nervous system. Welcome to your electromyogram (or EMG)! Connect your SpikerBox to your recording device and open up SpikeRecorder so you can visualize the spikes on your screen.

Familiarize yourself with what the signal looks like normally, with no activity, and then pick up something heavy. You will see and hear a difference in the signal on your screen and from the speaker: the "whoosh" is the sound of many action potentials firing in your muscles as the muscles contract. You are listening in on the conversation between your brain and your muscles! But are these really neurons we are recording from? We can zoom in and get some clues.

Experiment: Motor Unit Action Potentials

We've seen from our cockroach experiments that the individual spikes from neurons have very short peaks that are about 1ms wide. Since the neurons are fairly conserved between roaches and us humans, we can take a look at our signal to determine whether our spike widths are similar. Set up your recording file and zoom out so you can see about 15s of data, then start flexing your muscles.

We can see little "spindles" of activity. These are called spindles as each one sort of looks like a spindle of yarn. Let's zoom into a bit more to see what these spindles look like 10x closer in time.

0.1 s

We are starting to see some distinct spiking inside this burst of activity. Could these be neurons? Let's zoom in a bit more.

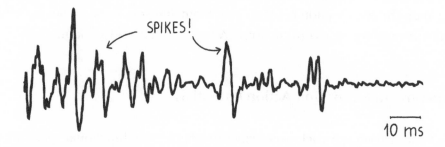

SPIKES!

10 ms

Examine these spike trains at this magnification and you'll see that something is different from your cockroach neurons. While the spikes from neurons were on the order of 1ms, the widths of these spikes seem to be much bigger, around 3–5ms. We can determine the average by setting a threshold, and take a snapshot of the signal each time the voltage crosses above that line. If we do this for all threshold crossings, we will have a picture of each spike. We can average them all together to determine the average waveform of our spikes.

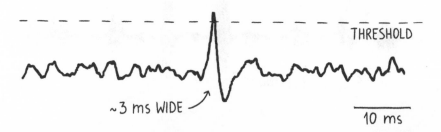

THRESHOLD

~3 ms WIDE

10 ms

We can now clearly see that the spikes from our muscle recording are indeed much bigger than our spikes from neurons. Behold! You have just discovered motor unit action potentials!

When your brain decides to move a muscle, neurons from your motor cortex (called "upper motor neurons") travel across your brain to your spinal cord, where they synapse with "lower motor neurons" (a.k.a. "alpha motor neurons"). These motor neurons then synapse with muscle to make a "motor unit." A motor unit is a single motor neuron and the multiple muscle fibers it innervates. A muscle fiber is a very special type of cell that can change its shape due to myosin/actin chains sliding across each other.

A single motor neuron can synapse with multiple muscle fibers. In general, a large, powerful muscle like your bicep has motor neurons that innervate thousands of muscle fibers, whereas small muscles that require a lot of precision, such as your eyeball muscles, have motor neurons that only innervate about 10 muscle fibers.

When a motor neuron fires an action potential, this causes a release of acetylcholine at the synapse between the neuron and the muscle (this type of synapse has a special name: the "neuromuscular junction"). This acetylcholine then causes changes in the electrical potential of the muscle. Once this electrical potential reaches a threshold, an actual action potential occurs in the muscle fiber! This action potential propagates across the membrane of the muscle, causing voltage-gated calcium channels to open, which begins the cellular cascade that ultimately causes muscle contraction.

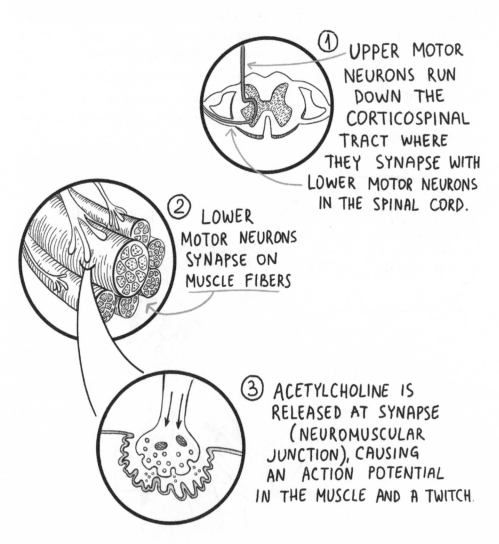

(1) UPPER MOTOR NEURONS RUN DOWN THE CORTICOSPINAL TRACT WHERE THEY SYNAPSE WITH LOWER MOTOR NEURONS IN THE SPINAL CORD.

(2) LOWER MOTOR NEURONS SYNAPSE ON MUSCLE FIBERS

(3) ACETYLCHOLINE IS RELEASED AT SYNAPSE (NEUROMUSCULAR JUNCTION), CAUSING AN ACTION POTENTIAL IN THE MUSCLE AND A TWITCH.

Since the sizes of the muscle fibers are much different from an axon, we record a much larger signal with wider spikes.

It's good to keep an idea of what the different electrophysiology signals look like. While the internal voltages in the cells of these signals are all similar, the way in which we record them makes the measured voltages very different. Since the voltage ranges may overlap, we can simply measure the spike width to determine whether that squiggly green line is a neuron, heart, or muscle.

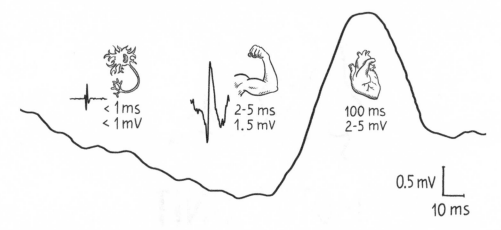

< 1 ms
< 1 mV

2-5 ms
1.5 mV

100 ms
2-5 mV

0.5 mV
10 ms

Follow-Up Questions

(1) You are recording through the skin and from several muscle fibers simultaneously. How would the recordings differ if you were recording right next to one of those cells? Inside the cell? Would you see the same number and type of spikes? How would the amplitude change? Think of a way you could test or compare this with invertebrate muscles, possibly using the Neuron SpikerBox.

(2) What caused the spikes that you saw? Specifically, what is occurring when the spike is positive? What is occurring when the spike is negative? Keep in mind that you are doing extracellular recordings, and of several fibers.

(3) Do you think you could record the activity of sensory neurons with this setup? The impulses are sent from where you sense something up to the brain. Try touching the hand from the arm you are recording, and see what happens!

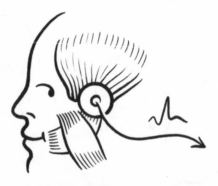

MOTOR UNIT
RECRUITMENT

19

Motor Unit Recruitment

Our bodies are masterful machines that are the envy of any roboticist. They are capable of many different complicated movements that we take for granted. You can easily reach for objects of different sizes and mass without noticing how difficult this is to accomplish. You can reach for a saltshaker, a coffee mug, or a full pitcher of water without thought. You can even pour the pitcher of water into an empty cup in your other hand without consciously noticing that one is getting lighter while the other is getting heavier . . . all while holding both the cup and pitcher completely still and even. This is a difficult job for a robot to accomplish.

We can also control our muscles very carefully. If you're a guitarist, you manipulate strings. If you're a surgeon, you manipulate a scalpel. If you're a scientist, you manipulate a pipette. You have been doing a fine motor action of turning the pages in this book. Each skillful action can be performed easily, often without paying attention. Every motor unit (a motor neuron in the spinal cord along with all the muscle fibers it innervates downstream) must be controlled carefully for us to walk, let alone dance gracefully. Motor units are the connection between the brain/spinal cord and the muscle. Here we will do some experiments to see how these motor units make precise movements possible.

Experiment: Muscle Recruitment in Chewing

In this experiment, we will eat some tasty snacks to learn how our brain carefully controls our motor units. Before we start, you'll need to gather some snack foods. Look for various types of textures and softness—for example, marshmallows, pretzels, gummy bears, beef jerky, or a vegetarian alternative such as really tough and chewy fruit leather. As best as possible, use food materials of roughly the same size. Once you have your food, let's make a hypothesis—a prediction about what we think we will observe in the experiment. For our hypothesis, we are going to rank several foods from largest to smallest expected EMG signal.

You will use the SpikerBox to record electromyograms of the masseter muscle, one of the two major jaw muscles responsible for chewing different foods. Place your electrode patches so that their centers (not the edges!) are lying about an inch apart on the masseter muscle. To figure out where that is, simply bite your teeth together gently . . . on the side of your jaw you will feel a muscle pushing back out . . . that's the one!

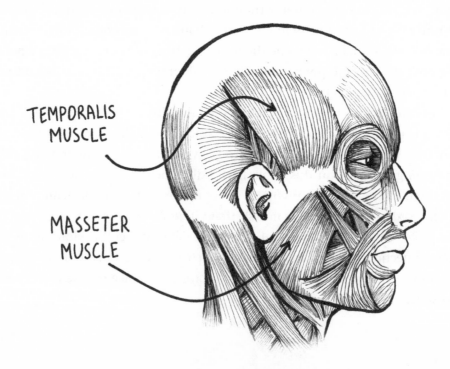

TEMPORALIS
MUSCLE

MASSETER
MUSCLE

Connect your recording (red) electrodes to the patches on the masseter muscle and the ground on the back of your wrist or mastoid. Connect the cables to the SpikerBox, and start your data recording and snacking! Do one snack at a time. Be sure to keep track of the time you started to chew and what type of food it was. Keep your chewing pace the same between foods. You will need this to make sense of the data. When you are all done, clean your hands and stop recording.

Let's take a look at what we recorded. Here is an example of a gummy bear:

To analyze these examples, we are going to measure just the chewing portion of the EMG. Use your device to measure the time length of just a single chewing epoch, and note the amplitude of the EMG. Find where the chewing burst begins and ends, and let's restrict our analysis to this section.

Compare your EMGs across the different foods. How close were you to your hypothesis?

MARSHMALLOW

BEEF JERKY

0.1s

The relative size of EMGs from traces of marshmallow and beef jerky reveal something interesting. It is important to note that a particular motor unit always has a similar size action potential. A larger EMG signal indicates that more motor units are being recruited by the brain. How did your brain know to produce a greater command for the beef jerky?

Each of your muscles is subdivided into functional groups of motor units. To achieve great things, like lifting a heavy weight, motor units are activated by the brain and join together in a systematic, predictable way to supply the force required to achieve strength. This teamwork among motor units is called "orderly recruitment." Motor units with the smallest number of muscle fibers begin contracting first during a movement, followed by the motor units with the largest number of fibers afterward, to allow for a smooth, strong muscle contraction.

If you look closely at the beef jerky EMG traces, you can see that the smaller motor units (that were sufficient for the marshmallow) engage early, only to quickly be joined by larger units. As you discovered, you don't have to think about recruiting motor units to create more force in your muscle. Your brain adjusts the neural output (the command) from the motor cortex, the motor units automatically respond, and you accomplish the movement.

Follow-Up Questions

(1) Try recording from using different parts of your mouth when chewing. Does the electrical signal look different? Why do you think it does or doesn't look different? Does it depend on what you're chewing on in that part of your mouth? Does chewing food normally require different parts of the mouth?

(2) We recorded from muscles responsible for clenching your jaw, but can you isolate which muscle is responsible for opening the jaw? Try the experiment again with foods that are sticky and that make it harder to open your mouth to chew again. Hint: the muscle is called the lateral pterygoid, and it can be hard to find.

(3) Try recording both the masseter and temporalis muscles at the same time with your SpikerBox. Do they show different activity when you are chewing?

EYE

MOVEMENTS

20

Eye Movements

Our brain is not limited to controlling only our body's position. It also controls two powerful cameras that capture the raw visual information from our surroundings—our eyes. Our eyes have a very important job. As visual creatures, we depend on our eyes to stream an incredible amount of information to let our brain identify objects, situations, and dangers. In this chapter, we will conduct experiments to explore how our brains control our eyes to move around and pay attention to important objects in the world.

Experiment: Horizontal Electrooculogram (EOG)

Take the headband we've been using for our EEG experiments, and place it just above your left eye. Be sure to adjust the headband such that the electrodes are positioned on both sides of the eye. Add some globs of electrode gel underneath the metal patches on the headband to make sure you have a good electrical connection to the skin. If you don't have a headband handy, you can use sticker electrodes in the same places.

Like all our electrophysiology experiments, we will need to add a ground to which all signals will be referenced. A good spot to add a ground patch electrode is on the bony protrusion behind your ear (the mastoid process), as it's known to be bioelectrically quiet. Now place the red alligator clips on the electrodes around the eye, and the black alligator clip on the ground behind your ear. Note that each red clip can lie on any side of the eye. Their placement, left or right, does not matter for these introductory experiments.

Plug the cable into your SpikerBox and start recording. If your data looks noisy, make sure your device isn't plugged into the wall, so that you can move a bit further away from any electrical outlets and fluorescent lights.

Now start moving your eyes. Try moving your eyes left and right, then up and down. Note that this experiment may be easier if a friend helps you to observe the data while your eyes are darting around. You will notice that moving your eyes left and right causes a much bigger deflection than moving your eyes up and down.

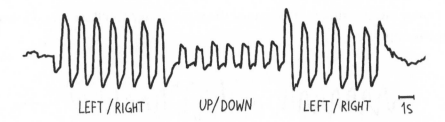

LEFT / RIGHT   UP / DOWN   LEFT / RIGHT   1S

Experiment: Vertical Electrooculogram (EOG)

Now adjust your headband so that one red recording electrode is just above your left eye, and place the second red electrode on a patch directly underneath the eye.

Repeat the experiment. Do you see the same thing? You'll see that it's actually the opposite.

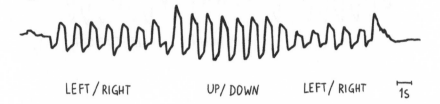

LEFT / RIGHT   UP / DOWN   LEFT / RIGHT   1s

Now the up-and-down movements are causing a slightly bigger signal. How could this be? What you just discovered is that your eye, much like your heart and brain, generates electrical potentials. The only difference is that this potential does not change quickly in the form of impulses like your heart and brain, but rather it produces a constant voltage difference between the front and back that we can measure. Specifically, the front of the eye (where the cornea is located) is more positive than the back of the eye (where the retina is).

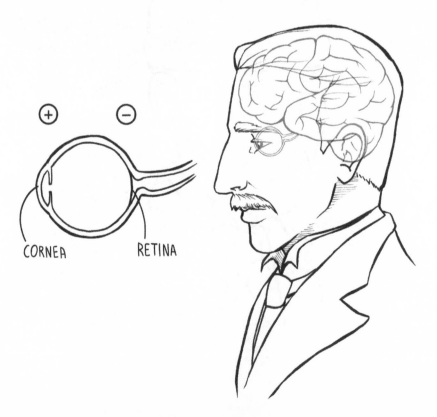

CORNEA   RETINA

Moving your eyes left and right swings the positive end closer to and away from the two horizontal electrodes, whereas the up-and-down movements tend to keep this charge at a relatively more constant distance. The opposite is true for the vertical electrode placement. Pivoting your eyes up and down will cause the positive charge to get closer and further to these positions, instead of the left-and-right movements. Recordings of this voltage are called an "electrooculogram" (EOG), and can be used to track the position of the eye.

Experiment: Saccades

Now that we know how to interpret the EOG, let's try another experiment. Set up your configuration for recording horizontal electrooculograms (on the left and right side of your eye). Start recording, but let's do more natural eye movements this time. Pick up a book or magazine and start to read. When you are all done, press stop, and take a look at the results.

1S

Here we see what looks like little repeating staircases. You may recognize the upward swing in voltage as a leftward movement. This is probably your eye moving to the left to read the next line—but what are the stairs?

While you may gracefully slide a finger across the page as you read these words, your eyes will make fast, jerky motions jumping along the lines of text. As you read this sentence, you are using the fovea of your eye. This is a little pit in the retina which contains closely packed cones, allowing you to determine that there is a capital A in this sentence. However, the fovea is very small, and it is not possible for it to look at the entire line at the same time. It's so small that we can only really see about 7–8 characters at a time. So your eyes must jump quickly across the text to process the information. These ballistic eye movements are called "saccades" (the French word for "to jerk").

Each saccade has a remarkably similar pattern. The voltage changes quickly as you jump to the next word, then is relatively constant for a short duration. This is when you are rapidly scanning the next piece of text with your fovea. Finally, the voltage swings upward when your eyes stop scanning and you jump to the next line.

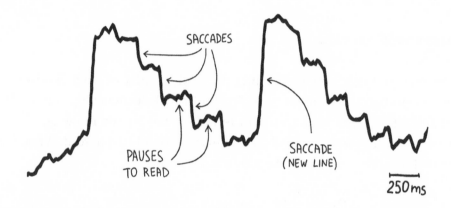

It's not just while reading that our eyes use saccades. Our eyes use these jerky motions to inspect all stationary objects. When we look at a beautiful painting, we shift our gaze from one point of interest to the next, taking in as much of the painting as possible. But our eyes are not just aimlessly wandering across the canvas. There is actually a lot of intent and purpose behind each movement. Our eyes are scanning the painting to find the most interesting parts. For your own experiment, record your saccades while inspecting a painting or a photo. How do they compare to those of reading a book?

Follow-Up Questions

(1) Do the potentials look different when your eyes are open or closed? Why?
(2) Is the amplitude of the potential affected by how far you move your eyes? How quickly? What other variables do you think could have an effect, and why?

(3) How does the placement of the electrodes affect the potential generated? Why does it have these effects? Think about the physics of how the potential is generated, and how electricity moves in the body.

(4) Try movements that require just your eyelids—closing your eyes (slowly), squinting, opening them wide. Why do we see or not see an EOG for these? If we do, how does it compare to the EOG seen earlier?

(5) Does whether or not someone needs glasses have an effect? Why or why not? If so, what is the effect?

MUSCLE FATIGUE

21
Muscle Fatigue

You're at your local gym, getting your pump on and lifting dumbbells. You're feeling strong and decide to try a 30 lb curl. Rep 1, rep 2, rep 3. . . . Ugh . . . why is it getting so hard to lift? In chapter 19, we saw the brain orderly recruiting motor units depending on the force needed. But what could be happening when you are lifting weights and your muscles start to burn?

The factors that explain fatigue are complex, and after more than 100 years of investigation, this is still a topic of active research! For example, short-term fatigue (the failure to do 30 lb curls, do more push-ups, etc.) is different from long-term fatigue (a marathon run, a 100-mile bicycle ride, or a full-day hike through the Rocky Mountains of Colorado). We can use our new skills of recording electromyograms to gain an understanding of why our muscles get tired.

Experiment: Isometric Biceps Hold

Place two patch electrodes on your biceps, and clip your red alligator recording leads to them. Place a patch electrode on the back of your hand and clip the black alligator ground lead to it. Then plug the electrode cable into your SpikerBox, and hook up your SpikerBox to your recording device.

Select a dumbbell that is at about 60% of your maximum lifting weight. Depending on your strength, this can be anywhere from 10–25 lbs (or ~5–12 kg). With your back to a wall to control your posture and arm position, bend your elbow at a 90-degree angle and hold the weight in your hand for as long as you can. Your muscles are working, but your joints are not moving. This is called an "isometric" (Greek for "same length") contraction. It may be possible that your wrist will tire faster than your biceps. If this happens, you can hang the weight off your wrist rather than holding the dumbbell in your hand.

Record your EMG during this task. Adjust the gain of your signal so that the signal doesn't get clipped on the screen of your recording software. Be sure you can see the peaks of the EMG spikes.

SAY NO TO FLAT TOPS!

Now that you are ready, pick up the weight and hold it steady for as long as you can. Observe the amplitude (height) and firing rate (number of impulses) in the signal. Make a note of when fatigue starts and when failure occurs. When finished, scroll through your data and see if you can spot any trends.

10 s

There does seem to be something going on here! The amplitude of the electromyogram signal is getting smaller the longer you hold the weight. Indeed, a closer look at the EMG reveals that the spikes from the larger motor units are systematically disappearing as the isometric hold goes on—until some point where the only units that are left are not enough . . . and you drop the weight. This reduction in the muscle's ability to produce the desired force or accomplish the desired movement is called "muscle fatigue."

Your brain can recruit new motor units to replace an already active motor unit that is experiencing fatigue. But the spare resources are limited.

To understand why motor units get tired, we have to look at the mechanism behind muscle movement. When a muscle cell fires an action potential, this causes a release of calcium (Ca^{2+}) inside the muscle fiber from the sarcoplasmic reticulum. The Ca^{2+} then flows into the sarcomere, which contains actin and myosin. This initiates a complex cellular reaction that allows the myosin to pull on the actin. The movement of myosin pulling on actin in the sarcomeres is called a "sliding filament model," and it consists of four steps.

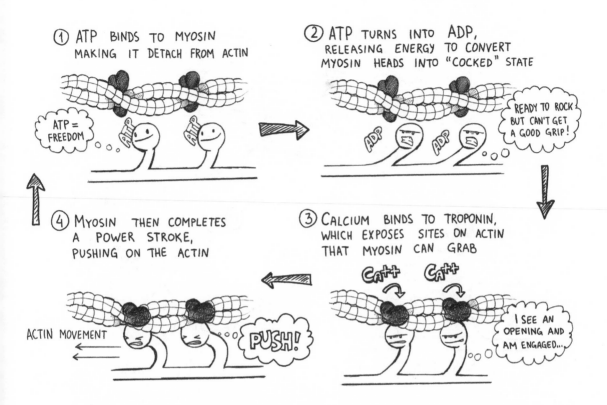

① ATP BINDS TO MYOSIN MAKING IT DETACH FROM ACTIN

ATP = FREEDOM

② ATP TURNS INTO ADP, RELEASING ENERGY TO CONVERT MYOSIN HEADS INTO "COCKED" STATE

READY TO ROCK BUT CAN'T GET A GOOD GRIP!

④ MYOSIN THEN COMPLETES A POWER STROKE, PUSHING ON THE ACTIN

ACTIN MOVEMENT

PUSH!

③ CALCIUM BINDS TO TROPONIN, WHICH EXPOSES SITES ON ACTIN THAT MYOSIN CAN GRAB

Ca++ Ca++

I SEE AN OPENING AND AM ENGAGED...

ATP is a small molecule that contains chemical energy. It is produced during the breakdown of food during energy metabolism. Oxygen, carried by the blood and delivered to the muscles, is needed to produce ATP. As long as oxygen is present and can be readily transported to the muscle cell, ATP can be produced at incredible rates. This is called "aerobic" contraction, meaning "using oxygen." However, contracting muscles can restrict blood flow and thus oxygen availability. Muscles could simply be working so intensely (a sprint at top speed) that there is not enough oxygen to meet the demand.

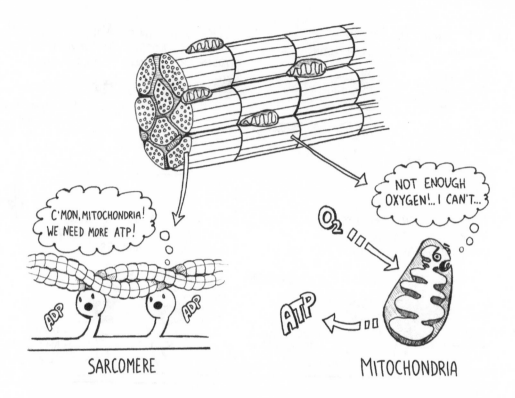

SARCOMERE   MITOCHONDRIA

If oxygen isn't available as an electron acceptor, the Krebs cycle and electron transport chain cannot operate, and the muscle must gain ATP from other sources. For example, for rapid, intense activity, phosphocreatine (synthesized from amino acids) can serve as a phosphate donor to allow ATP formation. This is called "anaerobic" contraction, meaning "not using oxygen."

Experiment: Modeling Rates of Fatigue in Isometric Grip

Let's add some data on the amount of force we can generate. To do that, we can use a hand-dynamometer or hand gripper with a strain gauge sensor attached. A gripper that can measure up to a 100 lb (45 kg) range is good enough.

The hand-dynamometer is designed to measure your force grip. Place two patch electrodes along your inner forearm. Connect the electrode leads and cables the same way you did for the bicep isometric hold. Only this time, you should be sure to manually keep track of the force data every 10s, or plug the gripper into the SpikerBox.

When you are ready, press record and start squeezing as hard as you can for as long as you can. You will see your forearm tighten up as your sensor starts spitting out your grip strength. Try to keep squeezing as hard as you can. You may notice that you will get little jumps in the force as you consciously reengage in keeping the force maximized. That's OK. Just keep applying maximum force until you give up. Whew! Press stop, and take a look at the recording.

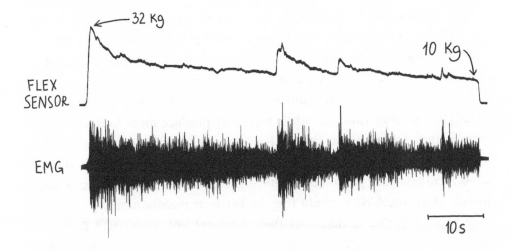

Here we can see both the strain gauge grip force as well as the EMG. We can take the first 5 seconds of the grip and the last 5 seconds of the grip to get an idea of how the EMG and force are changing. We can use this to estimate the rate in which our motor units are dropping out by fitting it to the linear function:

$$y = mx + b$$

Here "y" is the power of the EMG, "x" is time, "m" is equal to our EMG slope (rate of fatigue), and "b" represents the offset from the "y" axis. You can calculate the rate of fatigue using the equation:

$$m = \frac{(y_2 - y_1)}{(x_2 - x_1)}$$

Selecting the first 5s of the EMG trace, we can measure the power at 6.83 mV. The last 5s of the trace has dropped to 2.1 mV. We can also see that the entire flex was 116s. So our rate of fatigue becomes: m = (2.1 mV–6.83 mV)/(116s–0s) = –0.04 mV/s, with a peak force at 32 kg.

Try collecting data from multiple people and calculate the average rates of fatigue. You may find that while men may produce more initial force, the women may fatigue more slowly. For example, women may have had an average rate of -0.06 mV/s, while the men you sampled may have come in at –0.11 mV/s. Your preliminary data suggests that women may have better muscle endurance! Why would they be better at maintaining their strength over a longer period of time? Are they stronger? Not necessarily! Endurance isn't really about strength. Are they tougher? Certainly. Are they just flat out better in terms of muscular endurance? Maybe!

Research studies have been looking at this difference and have found that women's endurance is often greater than men's. Where can we see this in a non-scientific setting though? Women's impressive endurance gives them an edge in rock climbing—this is a sport where it is common to see women in direct competition with men. Research is also being done on women's endurance running, and it looks like men and women are on equal footing there as well!

Follow-Up Questions

(1) Try the biceps and forearms fatigue tests on both arms and hands to see if you observe anything different. As you know, you have a dominant arm/hand (being left-handed versus right-handed). Is your dominant arm/hand stronger or more fatigue resistant than the other?

(2) How can two muscles that are about the same size be so different in their fatigue properties? We didn't cover it here, but you can read about slow-twitch and fast-twitch muscle fibers to learn more.

(3) Are there muscles that are very fatigue resistant? Can you think of some examples? Work out your biceps for a month at your school gym. Measure your fatigue time and EMG changes before the period of training and after the period of training. Be sure to use the same test load/force.

(4) When hiking your favorite trail (like the Wonderland Trail or Torres del Paine), you may find, even if you are not very fit, you can hike for 6–10 hours. However, if you tried to lift a 100 lb (45 kg) barbell repeatedly, you would soon get tired within 5–30 reps over a couple of minutes, depending on your athletic ability. Why is the time scale of fatigue so different in these two activities?

REFLEXES

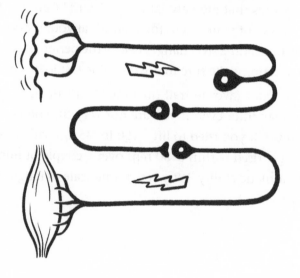

22

Reflexes

You don't have to think about reflexes. That's the whole point! They are very fast and they're automatic. Despite the magnificent information-processing power of the billions of neurons in our brain, we do need a lot of stuff to be done automatically. Reflexes move our body for us while our lumbering brains are still trying to process what happened.

Reflexes also take a cognitive load off of the brain. Without them, our conscious brains would be overloaded. Just imagine constantly thinking about how to position your body to keep yourself upright. When would you have time to engage in complex thoughts (black holes, neuroscience, what to do this weekend)? The original theory of how reflexes work came from none other than René Descartes.

I'm hit, therefore I kick!

RENÉ DESCARTES
1596 – 1650

When Descartes was a young boy living in France, long before he became a famous philosopher, he had an experience in the gardens of the French Royal Palace St. Germain that sparked his imagination. The French king was a bit of a prankster: he had set up lifelike mannequins that would jump out and surprise garden *flaneurs*, along with other mechanical automata such as statues that would retreat as art lovers attempted to get close enough to see more detail. René was startled, literally, by these mannequins that displayed a very lifelike action. Being curious, he investigated deeper and found that the motion was caused by a hydraulic system of water pipes.

Descartes writes of this experience in his 1633 essay "Treatise of Man," which details his theory of reflexes. René was an anatomist, and during his dissections he discovered that our nerves appeared to be small fluid-filled tubes (microscopes with enough magnification wouldn't be invented for a few more decades). He suspected that, much like the automated hydraulic exhibits in the garden of St. Germain, the nervous system was controlled by "animal spirits" (fluids finer than water), which would flow through the nerves of animals and people, giving rise to automatic responses.

He deals directly with some well-known reflexes. In one illustration, a small boy is placing his foot in a fire. René reasons that the particles of fire have enough force to displace the skin slightly and cause the nerve thread to act in the brain (as pulling on one end of a cord can ring a bell hanging on the opposite end). This interaction causes the animal spirits in the brain to enter the nerve, and move the muscles that withdraw his foot from the fire.

While we have seen that it is electricity not fluids that convey messages down axons, René's hypothesis that the brain is directly responsible for reflexes seems plausible. We do have axons that run up and down the spinal cord to and from the brain. We can do an experiment to test this out, but not by burning your foot in a fire . . . we will use a less painful stimulus.

Experiment: Patellar Stretch Reflex

For this setup, you will need a reflex hammer and an accelerometer. Connect an accelerometer to the SpikerBox and attach the sensor so that we can measure the acceleration of the hammer swinging. Have your subject sit on a table so that their leg can freely dangle, and contract their quadriceps (knee extensor) muscle so you can place two recording electrodes on either side of the muscles that are flexing above the knee. Place one on the vastus medialis, and the other on vastus lateralis. You can place your ground electrode on the back of your hand placed by the side of your leg.

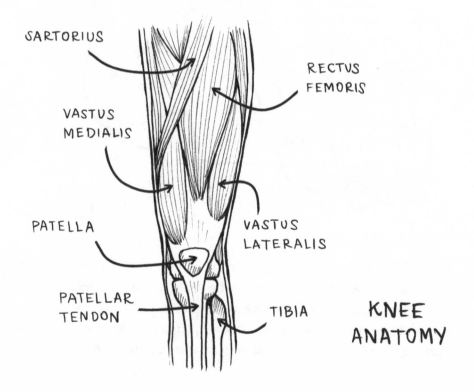

SARTORIUS

RECTUS FEMORIS

VASTUS MEDIALIS

PATELLA

VASTUS LATERALIS

PATELLAR TENDON

TIBIA

KNEE ANATOMY

While recording the EMG from the leg, swing the hammer so that it makes a striking contact with the patellar tendon (like your doctor does during your physical). See how long it takes from the contact of the hammer (indicated by a sharp change in acceleration) for the leg to generate an EMG response to kick the leg. Do that a few times until you can elicit a response with an EMG that you can see.

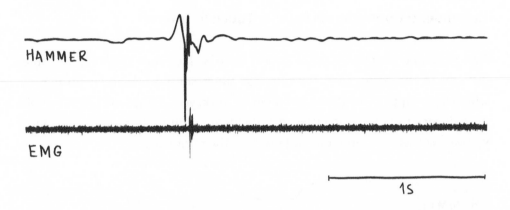

It happens pretty quickly, so you might have to zoom in to see it. Once you get close enough, measure the distance from the first peak in the hammer to the first peak in the EMG and write it down. We will call this the "response time," meaning how long it takes for the EMG to respond to the hammer.

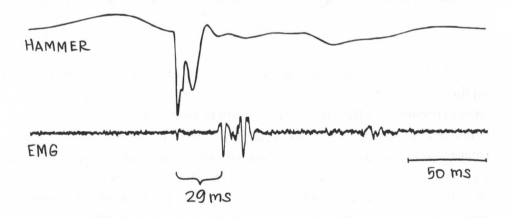

In this example, it took about 29ms from the hammer hitting the leg for the EMG to begin to respond. Now let's see if René was right. Does the signal go all the way up to the brain and come back down to the leg? What would happen if the opposite leg was hit with the hammer, while consciously kicking the same leg you recorded from in this experiment?

Experiment: Contralateral Patellar Touch Response

Here we are making the hypothesis that since both of your knees are the same distance to the brain, then the response times should be similar no matter which leg is tapped. For this experiment to work, your subject will need to voluntarily kick their leg when they feel the hammer hit their opposite leg. Once you've recorded this, zoom in and measure the response time.

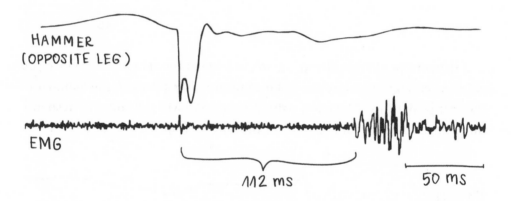

Woah! It takes almost four times longer for the EMG to appear when you hit the opposite leg compared to when you hit the patellar tendon. This would seem to disprove our hypothesis. What could be happening?

It turns out that Descartes was slightly wrong in his neuroanatomy. The patellar tendon has sensors that interact with the spinal cord through a reflex pathway. It doesn't need to go all the way up to the brain. Stretching the muscle with the hammer activates the muscle spindle at the end of the sensory neuron (embedded in your muscle) and starts the reflex. This reflex is to prevent overstretching of the muscle, and it compensates with a contraction.

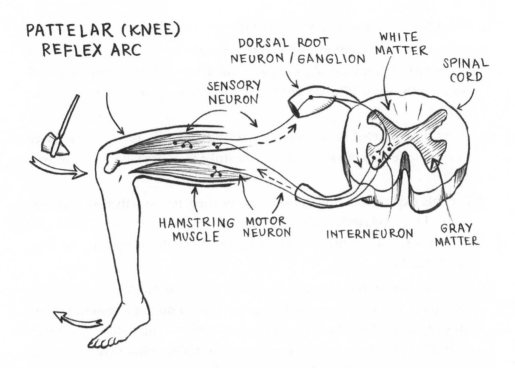

PATTELAR (KNEE) REFLEX ARC

SENSORY NEURON

DORSAL ROOT NEURON / GANGLION

WHITE MATTER

SPINAL CORD

HAMSTRING MUSCLE

MOTOR NEURON

INTERNEURON

GRAY MATTER

As you can see, there is only one connection (a synapse) needed for the information from the sensory neuron to get to the motor neuron and cause a muscle contraction. Because of this single synapse, this can happen very fast. In a young, healthy person, it takes 15–30ms for the stretch stimulus to produce a muscle contraction. This is super useful for correcting your muscle length in response to rapid changes such as a slip or trip. These situations require very fast corrections to prevent falling and injury. If you had to consciously flex your leg in response to an unexpected leg stretch while losing your footing, you would most likely fall!

Follow-Up Questions

(1) Why is the comparison between the two leg taps not very fair to Descartes? Are they both reflexes?

(2) What are some other reflexes you think you could test?

(3) Does the speed/amplitude of the response change depending on the state someone is in (for example, if they are tired, have a lot of energy, had coffee recently, etc.)? Why do you think these factors do or don't affect the reflex?

(4) Don't hurt yourself or anyone else investigating this, but does the speed or amplitude depend on the force you strike the knee with? Why or why not?

(5) For this experiment, the person having their reflex activated normally keeps the leg relaxed, but what if they try to resist the reflex? Does that affect the speed or amplitude at all, or does the reflex even happen? Why might this be?

(6) We have noticed two things that deserve further examination. The reflex time tends not to vary much in a single person during a session (always plus or minus 2ms), but the reflex time between individuals can range from 13–35ms. Since this is an unconscious reflex, what may be the reasons for such variability between individuals? Conversely, the reaction component tends to be more varied from trial to trial for a person (from 110–300ms). Why would this be?

(7) The knee-jerk reflex is the most famous example among a vast number of reflexes that all make our lives a little bit easier, because there is no thinking involved. Can you think of (or look up) five more reflexes of the human body? Can you design an experiment to investigate whether these are one-synapse (monosynaptic) reflexes?

REACTION TIME

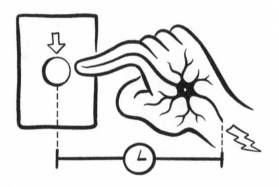

23

Reaction Time

In the world of sports, reaction time is extremely important.

Athletes need to be able to respond as quickly as possible. The fact that Usain Bolt can run 100 meters in 9.58 seconds is quite impressive. But it's even more impressive when you consider that this time includes a lot of information processing by his nervous system.

Let's look at the start of the race a bit more carefully. Sound energy from the starting pistol must enter his ears; hair cell neurons in his ear must vibrate and send information to the brain; neurons in the brain must identify the sound as the "go" signal, then motor commands must be sent down to his muscles. This whole process takes time, and this time from the beginning of the start signal to the start of the motor movement is called our "reaction time." But how long is the reaction time? We can do a simple experiment with a 12 in. (30 cm) ruler and a friend to find out.

Experiment: Simple Reaction Time

The setup is very easy. Have your friend sit at a table with their dominant hand exposed over the edge, and have them make a pinching hold with their fingers. Grab a hold of the ruler by pinching it on the short side near the

30 cm mark and hold it upright such that the 0 cm end is just between your friend's fingers.

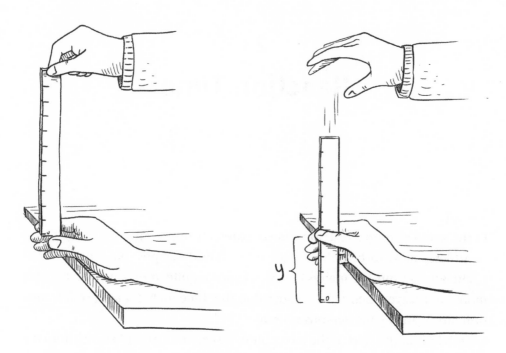

Tell your friend that when they see you release the ruler, they are to pinch shut and grab it as fast as possible. Try not to make any sounds, gestures, or other hints that you are releasing the ruler. They have to react to the visual stimulus of seeing the ruler being released. Record the centimeter mark where they pinched the ruler, and repeat this a few times.

Once you have the measurements, we just need to do some math. The formula below consists of three variables: y = the distance you measured in centimeters; g_0 = the acceleration due to gravity constant (981 cm/sec^2); and t = time in seconds.

$$y = \frac{1}{2} g_0 t^2$$

Since we are looking for the time (t), we can rearrange the equation using some algebra to look like:

$$t = \sqrt{\frac{2y}{g_0}}$$

Let's do some math.

SUPPOSE I RECORDED A MEASUREMENT OF 10 cm.

$$t = \sqrt{\frac{2 \cdot 10 \text{ cm}}{981 \frac{\text{cm}}{\text{sec}^2}}}$$

1) $\sqrt{\dfrac{20 \text{ cm}}{981 \frac{\text{cm}}{\text{sec}^2}}}$ I START BY MULTIPLYING A NUMERATOR. THIS EQUALS 20 cm. I WILL DIVIDE 20 BY 981 SO THE cm UNIT CANCELS OUT LEAVING ME WITH SEC SQUARED.

2) $\sqrt{0.02038736 \text{ sec}^2}$ I TAKE THE SQUARE ROOT OF THE NUMBER WHICH CANCELS OUT THE EXPONENT, GIVING ME THE UNIT OF SECONDS. PERFECT!

3) $t = 0.1427$ sec. $\boxed{0.14 \text{ sec}}$ I ROUND MY ANSWER TO 0.14s. THIS MEANS THAT THE RULER FELL 10 cm IN 0.14s UNTIL I CAUGHT IT. THEREFORE I HAD A REACTION TIME OF 0.14 SEC.

How fast was your friend? The typical reaction time of humans is between 0.15s and 0.3s. In the 2016 Olympic Games, Usain Bolt had a reaction time of 0.155s.

Experiment: Choice Reaction Time

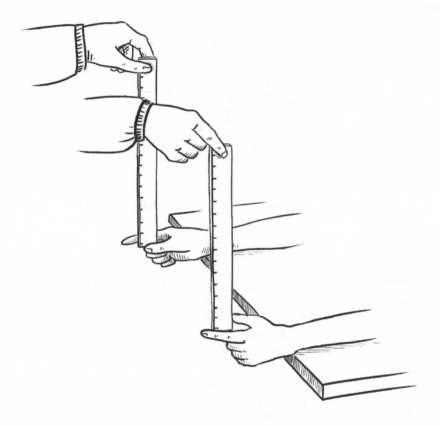

In this experiment, we will try to clock the speed of mental processing using simple rulers. You now know how fast, on average, your partner responds to seeing a ruler drop. Let's make it a bit harder. Let's force them to first make a decision. Have your partner sit at the table, like before, but now have them place both of their hands over the edge.

This time, you will suspend two rulers over their hands instead of just one. Each ruler needs to sit level at the 0 cm mark between their fingers. Tell your partner that you will release *just one* ruler, and they must pick the correct one and grab it as fast as possible. It is important that you tell them not to squeeze both hands, only one.

When you are ready to begin, randomly pick one ruler to drop. It does not matter which one: you will perform this test three more times, but never tell your partner which ruler you will drop. When finished, calculate the reaction time of this two-ruler choice task. Let's try to see how long the decision process takes. Since the first reaction time accounts for the visual to motor pathway, the difference between the two-ruler choice task and the one-ruler task will give you a rough estimate of how long the brain's decision-making process takes. Who would have thought that rulers could also be used to measure your internal mental cognitive abilities?

Follow-Up Questions

(1) Instead of visually watching the ruler drop, try having them close their eyes and listen to you saying "drop." Do you think the audio stimuli would have a faster reaction time on average? How about if they close their eyes and you tap them (a tactile stimulus)?

(2) Would you expect a difference in the average reaction times between men and women? What about a more athletic person compared to a more sedentary person?

(3) Do video-game players have quicker visual-motor reaction times due to the increased amount of screen-controller practice? Test this on a group of gamers and non-gamers.

(4) Why not test the tactile reaction time in the choice task? How could you redesign the experimental setup to test tactile reaction times in the choice task?

(5) As you know, you have a dominant versus a non-dominant hand. With only four trials, it is too hard to see a difference. Perhaps you should repeat the experiment 10–20 times with people who are right-hand and left-hand dominant to see if there is any difference between dominant and non-dominant hands.

CONCLUSION

24
Conclusion

You have reached the end (or perhaps just the beginning) of your journey with us as we explored your brain and the electrical signals of living things. We hope to have given you some ideas on how to frame your experiments and troubleshoot your signals, as well as some different ways to analyze your data. There is much more to discover, and there are many more unmarked paths to travel. For example, what about recording from the electrical signals of plants, or how about developing human interfaces where you can control devices with the electrophysiological signals of your body? You should now know enough that you could begin investigating these and other exciting fields on your own.

You may have noticed that our experiments dealt mainly with the inputs and outputs of the brain, recordings during sensory stimulation or motor outputs. Since neuroscientists can control and measure these inputs and outputs fairly accurately, we know a lot about how these functions work within the brain. But perhaps you picked up this book wanting to know how consciousness works—we (and many neuroscientists) would love to learn that too. How to measure conscious experiences can be difficult to measure. Some of the experiments in this book do begin to look at consciousness, such as the P300 and the readiness potential experiments. Perhaps you can think of creative ways to help push the boundary a bit further.

Our decision to build a book using an experiment-first over a theory-first approach was deliberate. The great biologist Sydney Brenner once said, "Progress in science depends on new techniques, new discoveries and new ideas, probably in that order." The goal of this book was to give you more intuition about how to ask questions using DIY techniques, how to analyze data to make discoveries, and how to weave those discoveries with what is known to gain more insights. And while the approach may have seemed a bit scattered (in one experiment recording the responses of moth antennae to odors, and in another experiment recording the change in heart rate as you place your hands in ice water), we hoped to give a broad overview of techniques to record how various electrical systems work together in your bodies and brains.

The title of this book, "How Your Brain Works," is indeed hubris. The honest truth is that there is still a lot to learn about how brains work, and we are still in the early stages of discovery. We chose this title to convey the notion that the brain is something we can understand in terms of principles and mechanisms. We hope that in reading, you are left with a belief system that the brain is not just a mysterious machine that controls our bodies and minds, but a system that can be studied and understood. Our goal is to provide you with the spark (Galvani's or Volta's) to make your own discoveries that reveal new things that were previously hidden.

We want to leave you with some advice to aspiring students of the brain. Science is a mix of curiosity, creativity, and action. We must nurture our curiosity to prepare our minds to observe something that perhaps was in the open all the time but that no one had studied (such as Hans Berger's discovery of the EEG). We must be creative to build models of what we think is happening inside our brains, or build instruments to measure and quantify our observations (such as a way to measure conduction velocity in earthworm nerves). Finally, we must also be people of action, conducting experiments or inventing new techniques and sharing them with the world. With these three qualities in mind, you will have the tools to make an impact in the field of neuroscience, and hopefully create new insights that will change the course of humanity.

So let's go! The next step is up to you. We hope you will take our advice to heart (and of course brain) and share your discoveries with us!

ACKNOWLEDGMENTS

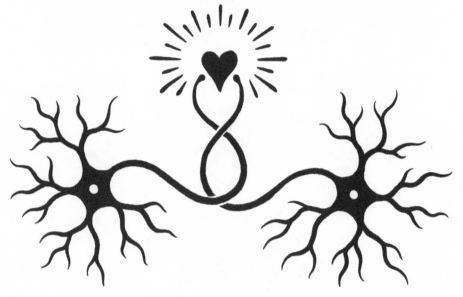

Acknowledgments

This book is really a story about Backyard Brains, the company we started as neuroscience graduate students at the University of Michigan.

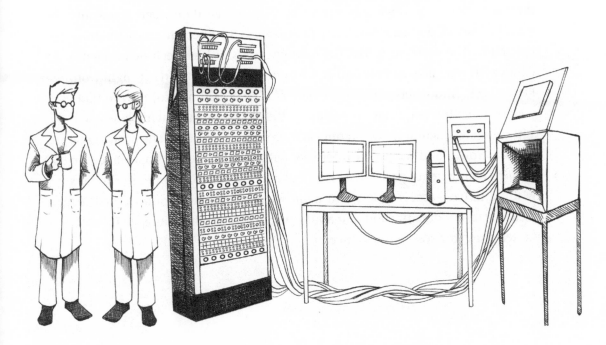

The experiments began with our cockroach-leg preparation, and a vision to reinvent how neuroscience is taught in schools. Over the past decade, we have developed our technology to handle a much wider and diverse set of signals, allowing for a deeper understanding of the nervous system.

The Backyard Brains project started as a poster at the Society for Neuroscience (SfN) about our SpikerBox. The poster was entitled "The $100 Spike" and was a self-induced engineering challenge to record a spike with equipment that would cost less than $100 to make. We submitted the abstract in May 2008, which laid out our goal and invited people to stop by to see how we did. We then got to work over the summer for the November SfN conference of 30,000 neuroscientists. We had learned from previous posters that presenting a "low-cost single-channel bioamplifier for education" would not get the attention it deserved in a busy crowd, so we instead positioned it as the cure for the zombie apocalypse. The theory was solid. Should the zombie apocalypse happen, all your loved ones would be affected by a disease that changed their behavior . . . so obviously it must be a neurological disorder! You would want to research and cure this horrible disease, so you need to study the zombie brains. The problem is that the manufacturers of high-tech neuroscience equipment were also zombies, so you would have to steal parts at night from RadioShack. And when you did, we had laid out a roadmap for how to build your own neuroscience rigs. Importantly, the device we built actually worked! You could hear neural spikes on our early prototype's speaker. The poster grew quite a crowd, and led to an interview with *Nature* and an appearance on *Nature*'s neuroscience podcast. This stimulated interest from other researchers wishing to have access to low-cost neuroscience tools, and soon we had the idea to start our organization to manufacture and research the use of DIY tools for teaching low-cost neuroscience.

We owe a debt of gratitude to the early mentors who encouraged us and helped get us training on how to start a company. Tim Marzullo received a Kauffman Foundation award designed for post-docs to commercialize their graduate school work. We learned how to pitch to investors, but soon found out that investors were not interested in an educational neuroscience start-up (no market existed). We made our website regardless, and had some early sales to the University of California at San Diego graduate students and other academics. But this venture would have probably been a short-lived side-hustle, as it was impossible to devote our time to develop it further without research funding. But there was hope. We learned of the Small Business Innovation Research (SBIR) award mechanism from the US government that was designed to solve just this problem. The SBIR funds R&D to develop technologies that are critical to the priorities of government agencies but that may not be attractive to standard investors. We developed a grant for the National Institutes of Health (NIH), and when it was funded, our lives changed, and our dream of Backyard Brains could survive. We could now devote 100% of our time to building hardware, software, and lesson plans. We would like to thank our National Institute of Mental Health (NIMH) program officer, Margaret Grabb, for all her support and for believing in two graduate students with a box of cockroaches. This book in fact was our final "specific aim," which was to publish a lesson book based on our experiments developed through our research grant. We are proud to say that the research reported in this book was supported by the NIMH under our SBIR Grant number MH093334: "Backyard Brains: Bringing Neurophysiology into Secondary Schools." We also thank Kris Bergman at the BBC Entrepreneurial Training & Consulting for all her advice on grant writing and for helping us navigate the complex world of government processes; and our PhD advisor, Daryl R. Kipke, for encouraging us to "run with this idea" as we were finishing our graduate work and for providing advice on our initial grant applications and pitches.

Greg would like to thank the TED organization, specifically Tom Riley, Logan McClure Davda, Shoham Arad, and the rest of the TED Fellows team for giving him the encouragement, mentoring support, and a stage to allow our ideas to percolate in the minds of others. Ideas are fragile—especially in the beginning formative state—and the TED organization's nurturing allowed our concept to blossom. The TED Talks helped our organization grow and reach our core audience of curious-minded learners.

Many of the experiments shown in this book were in collaboration with other researchers. Dr. Brian L. Tracy is a muscle physiologist at Colorado State University and has been a longtime colleague helping with all our EMG experiments. His student Breonna E. Holland developed the chewing experiment in chapter 19. Dr. Kenneth A. Norman and Dr. James W. Antony developed the TMR experiment in chapter 13, along with their high school intern, Robert Zhang, who programmed the memory game app. While a lab mate in grad school, Colin Stoetzner suggested the music stimulation experiment of chapter 5.

The hand-drawn illustrations first started with our graduate school friend Cristina Mezuk, who has a wonderful style that we continue to use to date. Over the years, we have added a few other artists to our team, including Maria Baykova and Matteo Farinella—the latter happens to be a neuroscientist and amazing author himself. All three of these artists have contributed to the illustrations in this book. Fonts were hand drawn by Matteo, Maria, and Aleksandra Gage.

We thank our many coworkers for their tireless devotion to making this book possible. Stanislav Mircic is one of the best engineers ever born, although he is too modest to admit it. He developed all our software and would cross over and quickly solve challenging hardware issues too. Will Wharton and Caitlin Clayton helped us review, edit, and organize early chapter drafts. Miroslav Nestorovic helped us organize the book development and coordinated with us authors, our publisher, and artists to get all the missing pieces in place. Jelena Ciric carefully reviewed each draft and added comments and suggestions. Data is a cornerstone of this book, and many were involved in collecting it. We thank Wenbo Gong for providing our EMG data in chapter 21, Will Wharton for the adaptation data in chapter 7, Zach Reining for collecting the ERG data in chapter 8, and Etienne Serbe-Kamp and Dan Pollak for developing the ERG vision experiment. Both Wenbo Gong and Ken Gage provided EKG data for chapters 16 and 18, respectively.

High school and undergraduate students have been instrumental in developing creative experiments, and we are thankful to all who worked long hours to build our DIY techniques. Our earthworm prep (chapter 4) was developed with Kyle M. Shannon, while Jess Breda developed the moth pheromone project in chapter 6. Ariyana Miri collected data for the EOG and saccades (chapter 20), and Joud Mar'i conducted the sleep and target memory reactivation experiments. The mu rhythm experiments (chapter 14) were conducted by Anusha Joshi. The P300 Response of chapter 13 was carried out by Kylie Smith. Maria Gerdyman developed the meditation experiment of chapter 15. Her beautiful sketches in her lab notebook were the source material for many of that chapter's illustrations.

On a more personal note, Greg wishes to thank his wife, Aleksandra, for her patience and understanding during the stressful start-up phase as a young scientist and entrepreneur; his daughters, Lila and Jane, for being a true source of joy, laughter, and inspiration; his family, for putting up with him as a child; and his father, for teaching him the value of creativity and humor. Tim wishes to thank his parents, William Marzullo and Lynn Harkins, and his grandparents for teaching him to never be intimidated by machines or language.

TABLE of EXPERIMENTS

Table of Experiments

- Exp. 1: Recording Spikes
- Exp. 2: Somatosensory Neurons
- Exp. 3: Somatotopy
- Exp. 4: Rate Coding
- Exp. 5: Earthworm Conduction Velocity
- Exp. 6: DC Microstimulation
- Exp. 7: AC Microstimulation
- Exp. 8: Frequency Analysis of Microstimulation
- Exp. 9: Silkworm Moth Mating Behavior
- Exp. 10: Silk Moth Chemotaxis
- Exp. 11: Silk Moth Electroantennogram
- Exp. 12: Neural Adaptation
- Exp. 13: Electroretinogram (ERG)
- Exp. 14: Color Electroretinograms
- Exp. 15: Broad Spectrum Electroretinograms
- Exp. 16: Cricket Cercal System
- Exp. 17: Inhibitory Neurotransmitters
- Exp. 18: Excitatory Neurotransmitters

THE APPENDICES

APPENDIX 1

HOW TO CARE FOR COCKROACHES

Appendix 1: How to Care for Cockroaches

Cockroaches are wonderful creatures and a perfect model organism for recording neurons, so you want to make sure they can live out their days in relative comfort. But where do you start? Below is a crash course for how to take care of your humble cockroaches. Armed with this knowledge, you can make a happy, comfortable home in which your roaches can live and relax.

Cockroaches can live 2–3 years with proper care, and you can even generate a self-sustaining colony with a little maneuvering. Hatchlings take about 6–8 months to reach sexual maturity. All roaches, no matter what stage, can live in the same container.

We recommend that you buy small terrariums from a local pet store. You can use any plastic container you have on hand as well, just be sure to make air holes. There are many cockroach species to choose from, but we found that discoid cockroaches (*Blaberus discoidalis*) have a few advantages: (1) They cannot crawl on glass or plastic, so if your container is big enough, you don't even need a lid! (2) They are from South America, so the temperate climates inside houses and classrooms mean they move a bit slower than their American or European counterparts. (3) They tend not to be as stinky as the more popular *dubia* species.

Once you have a container, fill the bottom of the terrarium with soil. We use unfertilized potting soil we buy from the local hardware store for a dollar per five pounds, but honestly, dirt from outside your house will work too. The cockroaches enjoy burrowing under the soil. Throw in some toilet paper rolls, egg cartons, or wood scraps for them to play and hide in.

For food, we prefer to use lettuce, as it provides water and doesn't mold quickly. Carrot slices also work. You can also add supplements of dry cat food, as it contains a lot of protein, which will cause the cockroaches to grow faster—particularly when they are small. You will have to be careful of excessive mold growth on the cat food though. Cockroaches also particularly like slices of banana, red apples, and oranges, but, like the cat food, you will have to be much more observant of mold growth.

For a more hands-off approach, you can purchase powdered roach or cricket chow and some water crystals. These often come in large bags or tubs and are designed for people who breed insects; but they should last you a long, long time. Just add water to the crystals, and place them in a low-walled container on the floor of the terrarium. Add some powdered chow to a similar container, but keep it on the opposite side of the water so they don't mix (again, mold!)

Your cockroaches should be kept at 70–85°F. They can handle colder temperatures but will not grow very fast. Replace the food every week, and spray the cage with a water-misting bottle to ensure humidity and give them water to drink.

Enjoy your new low-maintenance friends! You can listen to their pitter-patter at nights when they are most active.

APPENDIX 2

HOW TO BUILD A SPIKERBOX

Appendix 2: How to Build a SpikerBox

So why do we even need a SpikerBox to listen to neurons, brains, and muscles? The body communicates using electrical potentials that have very low electric current (I). To visualize this activity, we need to use special electronic circuits that have high impedance (or resistance, R). A low current combined with high resistance can produce a voltage large enough for us to use (from Ohm's law, $V=R*I$). We can continue to amplify the signal in stages to be able to see and hear the neural signals. Let's take a look at how this is done.

The invention of the transistor (by Bell Labs in the 1940s) ushered in a big change for electronics. A transistor allows a signal to become amplified (louder) through the use of semi-conducting materials such as silicon. This invention is considered to be one of the greatest of the twentieth century, as it has allowed computers to get very small and very fast. In the SpikerBox, we use transistors to amplify the biosignals many, many times.

Below is an annotated circuit diagram (or schematic) of a SpikerBox designed for recording neurons from insects. Viewing a circuit diagram is a lot like viewing a subway map. The topology (connections) is important, while the geographical location (where the components lie on a printed circuit board) can be changed to make the topology easier to view.

Power Circuit

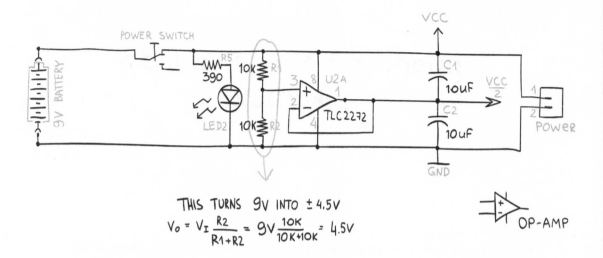

THIS TURNS 9V INTO ± 4.5V

$$V_0 = V_I \frac{R2}{R1+R2} = 9V \frac{10K}{10K+10K} = 4.5V$$

Operational amplifiers (op-amps) are symbolized by a right-pointing triangle and are used to amplify the difference between the input pins. Op-amps are made up of 20–30 transistors connected together to allow this special type of amplification. Typically, op-amps use dual-supply voltages (V+, V–), which allow the inputs and outputs to be referenced to ground (0 V). This would require two batteries (one each for V+ and V–). For portability, you can design the circuit to use a single 9V battery, and we therefore split the voltage into ±4.5 V using a voltage divider (R1/R2). The virtual ground is stabilized by an op-amp (Chip 2a) using a voltage follower. We used a TLC2272 as our op-amp, but similar parts could be used from other suppliers (TL074, OP291, OP293, or MCP602, to name a few).

Neural Amplification Circuit

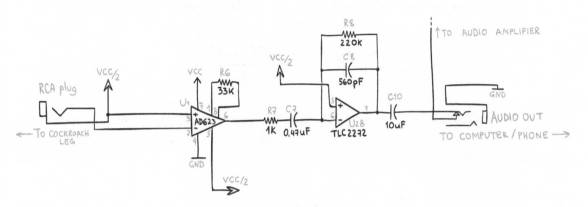

The two pins from the cockroach leg enter the circuit from the RCA plug on the left. This signal is amplified ~4× by a low-voltage instrumentation amp (AD623, Chip 1). An instrumentation amp is a special type of differential amplifier that is particularly suitable for use in measurement applications. You can also use other instrumentation chips too (e.g., INA118 from Texas Instruments).

STAGE 1 GAIN

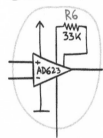

$$\text{gain} = 1 + \frac{100\,k\Omega}{R6} = 1 + \frac{100\,k\Omega}{33\,k\Omega} = 4.03\,x$$

STAGE 2 GAIN

$$\text{gain} = \frac{R8}{R7} = \frac{220K}{1K} = 220\,x$$

BUT IN REALITY THERE'S ROLL OFF, SO IT IS LOWER:

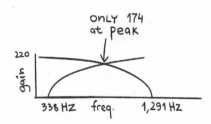

The gain (amplification) is set by adjusting the resistor across pins 1 and 8 according to the equation found in the AD623 data sheet:

$$\text{Gain} = 1 + \frac{100\,k\Omega}{R}$$

In our circuit, R = R6 = 33 kΩ, thus the gain is 4.03. The output of the AD623 is further amplified through a second TLC2272 op-amp (chip 2b), with the equation:

$$\frac{R_8}{R_7} = \frac{220\,k\Omega}{1\,k\Omega} = 220$$

This section of circuit also eliminates (filters out) frequencies that we are not interested in.

LOW-PASS FILTER

R8
—WW—
220K
C8
—)(—
560pF

$$f_c = \frac{1}{2\pi \cdot R8 \cdot C8} = \frac{1}{2\pi \cdot 220 \cdot 10^3 \cdot 560 \cdot 10^{-12}} = 1{,}291\ Hz$$

HIGH-PASS FILTER

R7   C7
—WW—)(—
1K   0,47uF

$$f_c = \frac{1}{2\pi \cdot R7 \cdot C7} = \frac{1}{2\pi \cdot 1 \cdot 10^3 \cdot 0.47 \cdot 10^{-6}} = 338\ Hz$$

The neural spike signals have frequencies of 300–1,300 in waveshape, so we designed the filter to allow those through. The resistor and capacitor in series

(C7 and R7) serve as a high-pass filter with a cut-off frequency determined by f=1/2πRC. Our high-pass cut-off is thus 338 Hz. The resistor and capacitor in parallel (C8 and R8) serve as the low-pass filter, with a cut-off frequency determined by the same equation, and is thus 1,291 Hz. The total gain of the circuit is 4.03*220 = 886.6. Note that these equations for gain and cut-off frequencies will work across any op-amp. This amplified signal then goes to the line out on the SpikerBox that can then be connected to a smartphone or computer using a smartphone cable (which goes into the microphone port for digitizing). You could also digitize the signal at this node and send over USB or Bluetooth. This is what we do on our commercial SpikerBoxes.

Audio Amplifier Circuit

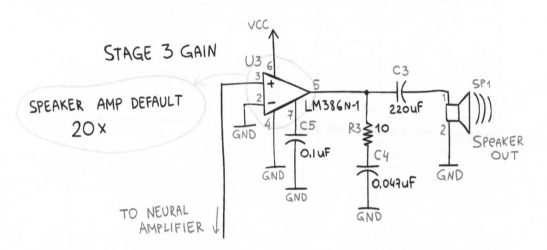

It's beautiful to hear the sound of spikes, so you will want to include an audio amplifier (using the LM386, chip 3) and a built-in speaker for listening to neural activity. The LM386 is configured to amplify the signal an additional 20x using the standard setup on the freely available LM386 data sheet from Texas Instruments.

Modify the SpikerBox to Record Other Signals

You can modify this SpikerBox circuit to work with other signals by adjusting the bandpass filter settings and gains. For example, if we want to record electromyograms (EMGs), we will need to change the bandpass to let through slower frequencies, down to 20 Hz. To do this, we need to change C7 (capacitor 7) to 6.8 uF. To only allow frequencies of EMG to pass through up to 500 Hz, we will change C8 to 18 nF (often written as 18,000 pF for some odd reason), and R8 to 18k. These are commonly available capacitor values; in this setup, the bandpass of the SpikerBox is now 20–500 Hz; the gain is now 4 × 18k/1k = 72. Below is a table that will allow you to modify the SpikerBox depending on the signal you are attempting to record:

INPUT SIGNAL			SPIKERBOX					
NAME	BANDPASS (Hz)	AMP(mV)	C7	R7	C8	R8	BANDPASS (Hz)	GAIN
SPIKES	300-1300	0.1-1	0.47uF	1k	560pF	220k	338-1291	900x
EMG	20-500	20-500	6.8uF	1k	18nF	18k	23-491	72x
EOG*	0-50	0.05-3.5	220uF	10k	2.8nF	1M	0.07-56	400x
EKG	0.05-100	0.1-10	220uF	10k	560pF	2.8M	0.07-101	1120x
EEG	0.1-100	0.01-0.1	220uF	1k	1.5nF	1M	0.72-106	4000x

*THIS RANGE SHOULD ALSO WORK FOR THE ANTENNA AND ERG EXPERIMENTS

If you don't want to build a SpikerBox yourself, we have prebuilt, open-source SpikerBoxes available for you to purchase at our website, backyardbrains.com. There, you can also find the SpikeRecorder app for collecting and analyzing your data.

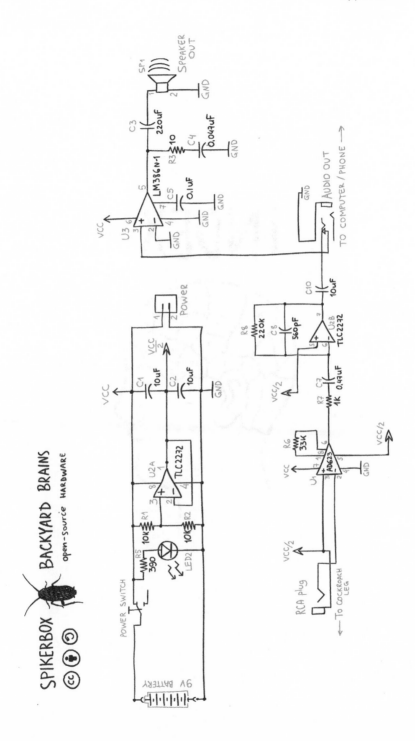

INDEX

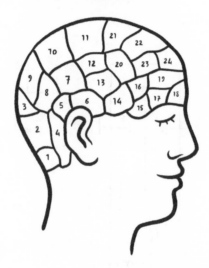

Index